Markus Lichtnecker

Determination of alpha_s via the Differential 2-Jet-Rate

Markus Lichtnecker

Determination of alpha_s via the Differential 2-Jet-Rate

with ATLAS at LHC

Südwestdeutscher Verlag für Hochschulschriften

Impressum/Imprint (nur für Deutschland/only for Germany)
Bibliografische Information der Deutschen Nationalbibliothek: Die Deutsche Nationalbibliothek verzeichnet diese Publikation in der Deutschen Nationalbibliografie; detaillierte bibliografische Daten sind im Internet über http://dnb.d-nb.de abrufbar.

Alle in diesem Buch genannten Marken und Produktnamen unterliegen warenzeichen-, marken- oder patentrechtlichem Schutz bzw. sind Warenzeichen oder eingetragene Warenzeichen der jeweiligen Inhaber. Die Wiedergabe von Marken, Produktnamen, Gebrauchsnamen, Handelsnamen, Warenbezeichnungen u.s.w. in diesem Werk berechtigt auch ohne besondere Kennzeichnung nicht zu der Annahme, dass solche Namen im Sinne der Warenzeichen- und Markenschutzgesetzgebung als frei zu betrachten wären und daher von jedermann benutzt werden dürften.

Verlag: Südwestdeutscher Verlag für Hochschulschriften GmbH & Co. KG
Dudweiler Landstr. 99, 66123 Saarbrücken, Deutschland
Telefon +49 681 37 20 271-1, Telefax +49 681 37 20 271-0
Email: info@svh-verlag.de

Zugl.: München, LMU, Diss., 2011

Herstellung in Deutschland:
Schaltungsdienst Lange o.H.G., Berlin
Books on Demand GmbH, Norderstedt
Reha GmbH, Saarbrücken
Amazon Distribution GmbH, Leipzig
ISBN: 978-3-8381-2861-0

Imprint (only for USA, GB)
Bibliographic information published by the Deutsche Nationalbibliothek: The Deutsche Nationalbibliothek lists this publication in the Deutsche Nationalbibliografie; detailed bibliographic data are available in the Internet at http://dnb.d-nb.de.

Any brand names and product names mentioned in this book are subject to trademark, brand or patent protection and are trademarks or registered trademarks of their respective holders. The use of brand names, product names, common names, trade names, product descriptions etc. even without a particular marking in this works is in no way to be construed to mean that such names may be regarded as unrestricted in respect of trademark and brand protection legislation and could thus be used by anyone.

Publisher: Südwestdeutscher Verlag für Hochschulschriften GmbH & Co. KG
Dudweiler Landstr. 99, 66123 Saarbrücken, Germany
Phone +49 681 37 20 271-1, Fax +49 681 37 20 271-0
Email: info@svh-verlag.de

Printed in the U.S.A.
Printed in the U.K. by (see last page)
ISBN: 978-3-8381-2861-0

Copyright © 2011 by the author and Südwestdeutscher Verlag für Hochschulschriften GmbH & Co. KG and licensors
All rights reserved. Saarbrücken 2011

It's just like flying a spaceship.
You push some buttons and see what happens.

Zapp Brannigan (FUTURAMA)

Have you tried turning it off and on again?

Roy Trenneman (The IT Crowd)

Abstract

The first determination of the strong coupling constant α_s via the differential 2-jet-rate in pp collisions at the LHC (at a center-of-mass-energy of 7 TeV) is presented. Data ($\int L\, dt = 700$ nb^{-1}) gathered by the ATLAS experiment are fitted by next-to-leading order (NLO) perturbative QCD predictions from calculations with the program NLOJET++. As an observable, the jet-flip-parameter from 3 to 2 reconstructed jets is investigated, using the infrared and collinear safe k_T jet algorithm in the exclusive reconstruction mode. The jet-flip-parameters from real data are compared to simulated data from Monte Carlo generators.

For the determination of α_s, real data have been corrected for the jet-energy-scale, whereas the calculations from NLOJET++ have been corrected for the influence of hadronization effects as well as the impact of the Underlying Event by applying bin-by-bin corrections. The fit between real data and the calculations from NLOJET++ yields a value of $\alpha_s(M_Z) = 0.120 \pm 0.001(stat.) \pm 0.005(syst.)$, which is in very good agreement with the current world average.

Zusammenfassung

In dieser Arbeit wird die erste Messung der starken Kopplungskonstanten α_s mithilfe der differentiellen 2-Jet-Rate bei pp Kollisionen am LHC (bei einer Schwerpunktsenergie von 7 TeV) vorgestellt. An Daten ($\int L\, dt = 700$ nb^{-1}) aus dem ATLAS Experiment werden dabei die Theorierechnungen in nächst-führender Ordnung (NLO) in der Störungsrechnung der QCD aus dem Programm NLOJET++ angepasst. Als Observable wird der Jet-Flip-Parameter untersucht, der den Übergang von 3 nach 2 rekonstruierten Jets beschreibt. Hierbei wird der infrarot- und kollinear-sichere k_T Jet Algorithmus im exklusiven Rekonstruktionsmodus verwendet. Die Jet-Flip-Parameter aus echten Daten werden mit simulierten Daten aus Monte Carlo Generatoren verglichen.

Für die Bestimmung von α_s werden einerseits die echten Daten um den Einfluss der Jet-Energie-Skala bereinigt und andererseits die Berechnungen aus NLOJET++ um den Einfluss der Hadronisierung und des Underlying Events korrigiert, indem die Einträge Bin für Bin korrigiert werden. Durch einen Fit zwischen echten Daten und den Berechnungen aus NLOJET++ ergibt sich ein Wert von $\alpha_s(M_Z) = 0,120 \pm 0,001(stat.) \pm 0,005(syst.)$, der sehr gut mit dem Weltmittelwert übereinstimmt.

Contents

1	**Introduction**	1
2	**Theory**	3
	2.1 Standard Model	3
	2.2 Quantum Chromo Dynamics	5
	2.2.1 Color-charge	6
	2.2.2 Strong Coupling Constant α_s	7
	2.3 Hadronization	8
	2.4 Determination of α_s	10
	2.5 Parton Distribution Functions	12
	2.6 Background Processes	13
	2.6.1 Minimum Bias	13
	2.6.2 Underlying Event	14
	2.6.3 Pile-up	16
3	**LHC and ATLAS**	17
	3.1 LHC	17
	3.2 ATLAS	18
	3.2.1 Coordinate System	19
	3.2.2 Magnet System	20
	3.2.3 Inner Detector	21
	3.2.4 Calorimeter	22
	3.2.5 Muon Spectrometer	22
	3.2.6 Trigger	23
	3.3 Data & Computing Grid	25
4	**Jets**	27
	4.1 Jet Production	27
	4.2 Cone Algorithm	27
	4.3 k_T Algorithm	29
	4.3.1 Inclusive Mode	30
	4.3.2 Exclusive Mode	30
	4.3.3 Anti-k_T Algorithm	31
	4.4 Inputs to Jet Reconstruction	32
	4.5 Jet Correction	33
	4.5.1 Correcting for Calorimeter Response	33
	4.5.2 Jet-Energy-Scale	33
	4.5.3 Jet Cleaning	34

5	Analysis Software	37
	5.1 NLOJET++	37
	5.2 PYTHIA and Underlying Event Models	38
	5.2.1 PYTHIA	39
	5.2.2 Underlying Event Models	40
	5.3 HERWIG	42
	5.4 ATHENA	45
6	Differential 2-Jet-Rate	47
	6.1 Motivation	47
	6.2 Studies with NLOJET++	47
	6.2.1 d_{23} Distributions of 2-Parton-Events	48
	6.2.2 d_{23} Distributions of 3-Parton-Events	50
	6.2.3 d_{23} Distributions of 4-Parton-Events	52
	6.2.4 Comparison between BORN and FULL "minus" NLO	54
	6.3 Real Data Analysis	56
	6.3.1 Datasets	56
	6.3.2 Jet Cleaning	58
	6.3.3 Separation from 4-Jet-Events	62
	6.4 Comparison to Simulations	64
	6.4.1 Comparison between Data and PYTHIA	64
	6.4.2 Comparison between Data and NLOJET++	67
	6.4.3 Comparison between PYTHIA and NLOJET++	68
7	Corrections	71
	7.1 Jet-Energy-Scale	71
	7.2 Hadronization	71
	7.3 Underlying Event	76
	7.4 Low-p_T Method	79
8	α_s-Fit and Systematic Uncertainties	87
	8.1 α_s-Fit	87
	8.1.1 LO α_s	88
	8.1.2 NLO α_s	89
	8.2 Systematic Uncertainties	91
	8.2.1 JES Uncertainty	91
	8.2.2 Impurity due to 4-Jet-Events	94
	8.2.3 Renormalization Scale Uncertainty	94
	8.2.4 PDF Uncertainty	95
	8.2.5 Hadronization Uncertainty	98
	8.2.6 Underlying Event Uncertainty	100
	8.3 Final Results	101
9	Summary	103
	Bibliography	105

Chapter 1

Introduction

Since ancient times matter and its structure have been investigated by mankind. Beginning with thought experiments of philosophers, the science led to the biggest experiments on earth. Elementary particle physics uses experiments with cosmic rays as well as huge particle accelerators, like the Large Hadron Collider (LHC) at CERN[1] near Geneva, Switzerland, to study the properties and interactions of matter. The LHC (and with it the ATLAS[2] experiment) started its operation in September 2009. It has been designed to collide protons at a center-of-mass energy[3] of $\sqrt{s} = 14$ TeV (currently, the LHC is operating at $\sqrt{s} = 7$ TeV) with a final instantaneous luminosity of up to $\mathscr{L} = 10^{34}$ cm^{-2}s^{-1}.

At such high energies, it is now possible to find (or exclude) predicted particles, which have not been observed yet due to their very large mass. Nonetheless, before new particles can be discovered, and thereby new theoretical models confirmed, a good understanding of the Standard Model (SM) at LHC scale is crucial. All known elementary particles and interaction forces (excluding the gravitation) are included in this powerful model: the electromagnetic, the weak and the strong interaction. The latter describes the force between quarks and gluons and has a range of about 10^{-15} m. The interaction force is conveyed by eight gluons, being discovered in 1979 via 3-jet-events at PETRA[4] at DESY[5]. The strength of the strong interaction is described by the strong coupling constant α_s. By combining many different measurements, the world average was set to $\alpha_s(M_Z) = 0.1184 \pm 0.0007$ with a Z boson mass of $M_Z = 91.1876 \pm 0.00021$ GeV (values are taken from [1]), which is about two orders of magnitudes above the electromagnetic force.

When starting a new experiment, first of all the detector has to be understood and it has to be shown that the experiment works well, reproduces the results from former colliders and is consistent with the theoretical extrapolation to the high collision energies of the LHC. Sophisticated technical innovations and improved analysis techniques make it possible to measure several properties of particles and their couplings with an accuracy and precision that are second to none [2].

A basic quantity for testing the Standard Model and especially Quantum Chromo Dynamics (QCD) is the strong coupling constant α_s, describing the strength gluons couple to colored particles. As α_s is not a constant - contrary to the misleading name - but varies with the transfer of momenta Q, it opens the opportunity to compare its value with former experiments and, in addition, to determine it in regions of Q not yet investigated.

As a test of QCD, this thesis deals with the determination of α_s via the ratio of 3-jet-events to 2-jet-events. The jets are reconstructed using the k_T jet algorithm in the exclusive mode, with the algorithm being forced to find 3 jets in the final state. The differential 2-jet-rate is measured via the jet-flip-values, describing the transition from 3 to 2 reconstructed jets.

---

[1] European Organization for Nuclear Research (french: Conseil Européen pour la Recherche Nucléaire)
[2] A Toroidal LHC AparatuS
[3] The *natural units* ($\hbar = 1$ and $c = 1$) are used throughout this thesis.
[4] Positron-Elektron-Tandem-Ring-Anlage
[5] Deutsches Elektronen-Synchrotron in Hamburg

The presented method has the advantage that it can be done with early data gathered by the ATLAS detector.

This thesis is divided into 9 chapters. The theory chapter describes the SM, the QCD, the hadronization process and the method used to determine α_s as well as the parton distribution functions. Finally, some background processes (such as the Underlying Event) are explained. In chapter 3, the LHC, the ATLAS detector and the data & computing grid are introduced, followed by a chapter about jets and jet algorithms. The analysis software used is presented in chapter 5. Chapter 6 motivates why the differential 2-jet-rate has been used to determine α_s and presents the distributions from calculations with NLOJET++. Then, real data is analyzed after applying jet cleaning cuts. The differential 2-jet-rate is compared to the differential 3-jet-rate before comparing real data to simulation. In chapter 7, the influence of the jet-energy-scale, hadronization effects and the UE are corrected for. α_s is then determined via fits to the differential 2-jet-rate distributions in chapter 8. Also some systematic uncertainties are investigated. Finally, chapter 9 summarizes the results.

Chapter 2

Theory

This chapter (based on [2-6]) focuses on the theoretical background of this thesis. First of all, the Standard Model (SM) of elementary particle physics is described, followed by an overview of Quantum Chromo Dynamics (QCD). Furthermore, the hadronization of quarks and gluons into color-neutral particles is introduced. Then, the method used to determine α_s is shown. After describing the structure of the proton, some background processes are explained with a focus on the Underlying Event (UE).

2.1 Standard Model

The Standard Model of particle physics comprises the known elementary particles and the interactions between them. It has passed (excluding the Higgs boson) all theoretical and experimental tests to a level smaller than 0.1%. Richard Feynman[1] already said that "the Standard Model is working too well".

According to the SM the whole matter consists of twelve fermions[2] (see table 2.1): six quarks[3] (up, down, strange, charm, bottom and top), which undergo the electroweak as well as the strong interaction, six leptons (electron, muon, tau and respective neutrinos), being solely subject to the electroweak interaction (because they don't carry color-charge) and the according antiparticles[4].

The fermions can be grouped in three generations, each with two leptons, two quarks and the corresponding antiparticles.

Quarks never appear as free particles, but always as color-neutral hadrons. These composed particles are either mesons, i.e. quark-antiquark-pairs, or baryons, consisting of three quarks, like the proton, comprising two up-quarks and one down-quark.

The interactions between the particles are represented by field quanta, commonly known as gauge bosons[5], which carry the force. The Feynman diagrams of some fundamental fermion-boson couplings in perturbation theory are shown in figure 2.1. The range of these bosons is - agreeable to the Heisenberg uncertainty principle - linked to their mass (see table 2.2) - except for the gluons, which transmit the strong interaction.

---
[1] Richard Phillips Feynman (1918-1988) was an American physicist, who won the Nobel prize in physics in 1965 for his contributions to Quantum Electro Dynamics (QED) [7].
[2] Fermions are spin-$\frac{1}{2}$-particles, obeying Pauli's exclusion principle. In addition, they adhere to the Fermi-Dirac-Statistics.
[3] The name quark has its origin in the book *Finnegans Wake* by James Joyce. The American physicist Murray Gell-Mann (born 1929) liked the sentence *Three quarks for Muster Mark!* so much that he adapted the name quark for these subatomic particles (only three quarks were known at that time) [7].
[4] Antiparticles have the same masses as the according particles, but opposite electrical charge, color and third component of the weak isospin [3].
[5] Bosons have a whole-number spin and obey to the Bose-Einstein-Statistics. They are not subject to Pauli's exclusion principle.

Quarks

Generation	Flavor	Symbol	Charge [e]	Mass [GeV]
1	up	u	$+\frac{2}{3}$	0.00225 ± 0.00075
	down	d	$-\frac{1}{3}$	0.005 ± 0.002
2	charm	c	$+\frac{2}{3}$	1.25 ± 0.09
	strange	s	$-\frac{1}{3}$	0.095 ± 0.025
3	top	t	$+\frac{2}{3}$	174.2 ± 3.3
	bottom	b	$-\frac{1}{3}$	4.2 ± 0.07

Leptons

Generation	Name	Symbol	Charge [e]	Mass [GeV]
1	electron	e	-1	511×10^{-6}
	electron neutrino	ν_e	0	$< 2.2 \times 10^{-9}$
2	muon	μ^-	-1	105.7×10^{-3}
	muon neutrino	ν_μ	0	$< 170 \times 10^{-6}$
3	tau	τ^-	-1	1.7777
	tau neutrino	ν_τ	0	$< 15.5 \times 10^{-3}$

Table 2.1: The Fermions at a glance [1]. All stable matter forming our visible universe is composed by particles of the first generation. Particles of the second and third generation (having higher masses) only have a short lifetime.

In this way, the electromagnetic force (described by the QED) has an infinite range, because it is conveyed by massless photons. Accordingly, the range of the weak interaction is rather small ($\ll 10^{-16}$ m), due to the large masses of the $W^\pm$ bosons[6] of 80 GeV and the Z^0 boson[7] of 91 GeV respectively. However, the range of the strong interaction is not infinite, although the eight gluons representing this force are massless. This can be traced back to the fact that gluons interact among each other (see section 2.2.1).

The forth fundamental force, gravity, is not part of the SM. The according particle, the graviton (with spin 2), has not been observed yet. In comparison to the other interactions, the gravitational force is almost negligible (in relation to the strong interaction it has merely a magnitude of 10^{-38}).

It is a great achievement of the Standard Model that electromagnetic and weak force could be joined in a common theoretical framework - the electroweak theory.

	EM Force	Weak Force	Strong Force
Strength	$7.30 \times 10^{-3} \approx \frac{1}{137}$	1.02×10^{-5}	≈ 1
Range [m]	∞	$\ll 10^{-16}$	$10^{-15} - 10^{-16}$
pertains to	charged particles	fermions	quarks
conveyed by	γ (photon)	$W^\pm$, Z^0 (gauge bosons)	g (gluon)
Mass [GeV]	0	$\approx 10^2$	0

Table 2.2: The elementary forces and their mediating gauge bosons of the Standard Model [5]

---

[6] $W^\pm$ bosons couple to weak isospin doublets of left-handed fermions and - as they carry an electrical charge - also to photons.

[7] The Z^0 boson acts on both left- and right-handed particles, but not on photons, as it is electrically neutral.

2.2. Quantum Chromo Dynamics

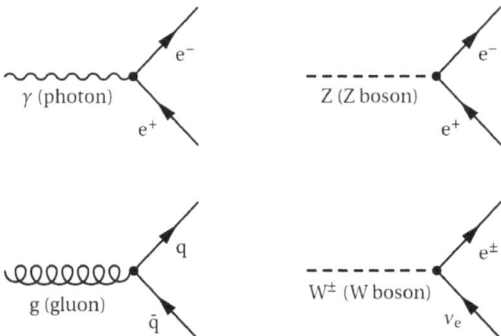

Figure 2.1: Feynman diagrams exemplifying some of the fundamental fermion-boson couplings in perturbation theory [5]

The reason for the masses of the W and Z bosons (discovered in 1983) is supposed to be described by the Higgs mechanism, i.e. a spontaneous breaking of the electroweak gauge symmetry. In this theory, (at least) one further scalar particle is needed: the Higgs boson (H), named after the Scottish physicist Sir Peter W. Higgs. The Higgs boson of a symmetry breaking background field couples to the W and Z bosons and provides them with an effective mass. For a further description of the Higgs mechanism, see e.g. [8]. In this theory, the fermions get their masses by Yukawa-couplings to the Higgs particle. The experimental proof of the Higgs boson is one of the main research goals of the LHC[8].

With the exception of the Higgs mechanism, the SM is based on the principle of gauge invariance, i.e. the invariance of a gauge field under local phase transformations, e.g. $\Phi(x) \to e^{i\Theta(x)}\Phi(x)$.
The symmetry group of the SM is a $U(1)_Y \otimes SU(2)_L \otimes SU(3)_C$ gauge symmetry. $SU(3)_C$ (the C represents the quantum number color) is the symmetry of the strong interaction and $U(1)_Y \otimes SU(2)_L$ (the Y corresponds to the weak hypercharge, the L to the isospin) the symmetry of the electroweak interaction. $U(1)_{em}$ as the symmetry of the electromagnetic interaction is a subgroup of the electroweak interaction.

2.2 Quantum Chromo Dynamics

In order to understand the strong interaction, the QCD[9] is needed, being formulated in 1973 [11]. In this quantum field theory, the strong interaction between quarks and gluons is described by a new charge, being based upon the electrical charge of QED.
This charge is called color[10], because the three occurring charges have been assigned to the colors red (r), green (g) and blue (b) (accordingly antired ($\bar{r}$), antigreen ($\bar{g}$) and antiblue ($\bar{b}$) for antiparticles).

---

[8]The predecessor-experiment LEP (Large Electron-Positron Collider) could only determine a minimum mass of 114.4 GeV [9]. LEP ran from 1989-2000 at a collision energy of up to 209 GeV [10].
[9]The name Quantum Chromo Dynamics comes from the Greek term *chromos* = color.
[10]The name color is just a matter of nomenclature in order to distinguish these quantum numbers and should not be misapprehended as an indicator that quarks are literally colored.

2.2.1 Color-charge

Quarks carry these color-charges[11] in addition to their electrical charge. Therefore, each quark flavor exists in three different colors.
In addition to the quarks also the eight gluons ($m = 0$ and $J^P = 1^-$) are color-charged (in contrast to QED, where the photons are electrically neutral). In the case of the gauge bosons, this charge is a combination of color and anticolor. Due to their color-charge, the gluons interact with themselves (this is called self-interaction of gluons), besides their coupling to quarks.
In accordance with the $SU(3)_C$ symmetry group, the $3 \otimes \bar{3}$ color combinations split into a color octet and a color singlet. The latter

$$\sqrt{\frac{1}{3}}(r\tilde{r} + g\tilde{g} + b\tilde{b}) \tag{2.1}$$

is invariant under rotations in the color space and hence color-neutral. Therefore only the octet states couple to color-charged particles. These color-states are:

$$r\tilde{g}, r\tilde{b}, g\tilde{b}, g\tilde{r}, b\tilde{r}, b\tilde{g}, \sqrt{\frac{1}{2}}(r\tilde{r} - g\tilde{g}), \sqrt{\frac{1}{6}}(r\tilde{r} + g\tilde{g} - 2b\tilde{b}) \ . \tag{2.2}$$

In figure 2.2 the fundamental Feynman diagrams of the strong interaction (including the self-coupling of the gluons) are shown.

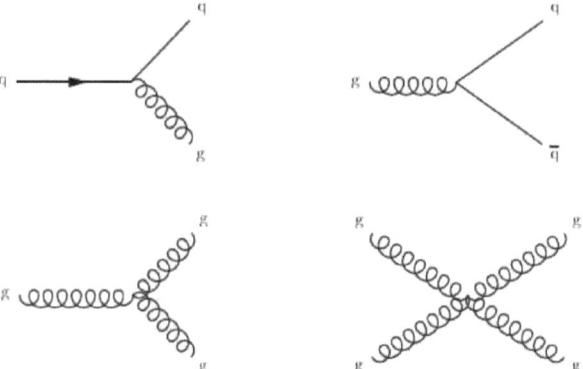

Figure 2.2: Feynman diagrams of fundamental QCD interactions. From top left to bottom right: gluon emission, gluon splitting up into a quark-antiquark-pair, gluon self-interaction of three and four gluons [6]

---

[11]Color-charges can be concluded experimentally by e.g. looking at the cross section ratio $\frac{e^+e^- \to hadrons}{e^+e^- \to \mu^+\mu^-}$. The quantum number color was required to preserve Pauli's exclusion principle when discovering the Ω^- particle (composed of three strange-quarks), and accordingly the Δ^{++} particle, consisting of three up-quarks.

2.2.2 Strong Coupling Constant α_s

The self-interaction of gluons is responsible for the variation of the interaction potential between quarks (or colored particles in general), which grows with increasing distance along the lines of a stretched spring.
The force that binds the particles together is described by the strong coupling constant α_s. Analogous to the finestructure constant of the QED, this force is defined as

$$\alpha_s = \frac{g_s^2}{4\pi} , \qquad (2.3)$$

where g_s represents the color-charge.
As already announced, α_s is not a real constant, because the value depends on the energy-scale Q and therefore on the distance of the color-charged particles to each other (see figure 2.3).

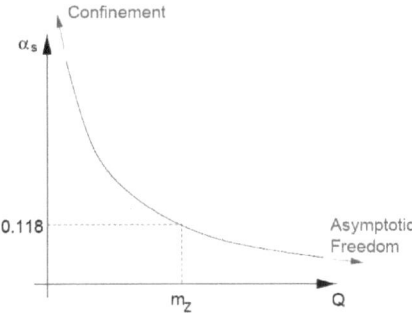

Figure 2.3: Q-dependency of the strong coupling constant α_s [6]

The running of α_s is explained by the vacuum polarization. In contrary to the naive picture, the vacuum is not empty, but has a complex structure. Therefore it gets polarized in presence of a color-charge and bare charges get shielded. Visible charges thus become energy- and distant-dependent. On the one hand, this leads to the *asymptotic freedom* when distances are small (and consequently transverse momenta transfers Q large):

$$\lim_{Q \to \infty} \alpha_s(Q) \to 0 . \qquad (2.4)$$

Quarks can consequently move virtually free within small distances, because of the small attraction forces. These regions can be handled by perturbation theory.
On the other hand, when going to large distances, the energy-density between particles becomes larger until quark-antiquark-pairs and gluons are produced from the vacuum, comparable to a paramagneticum for color-charges. This is energetically more favorable than enlarging the distance between the quarks.

$$\lim_{Q \to \Lambda_{QCD}} \alpha_s(Q) \to \infty , \qquad (2.5)$$

where Λ_{QCD} is the only free parameter of QCD with a value of a few hundred MeV.
Colored particles can hence never appear individually, but only as color-neutral hadrons. This fact is called *confinement*. The confinement is outside the regime of perturbation theory calculations applying Feynman diagrams [3, 12].

2.3 Hadronization

Partons (i.e. color-charged quarks and gluons) do not exist as free-propagating particles, in contrary to color-neutral hadrons. As perturbation theory cannot be used to calculate the involvement of partons (which can originate from the vacuum) into hadrons[12] because of the confinement, phenomenological models have to be implemented.

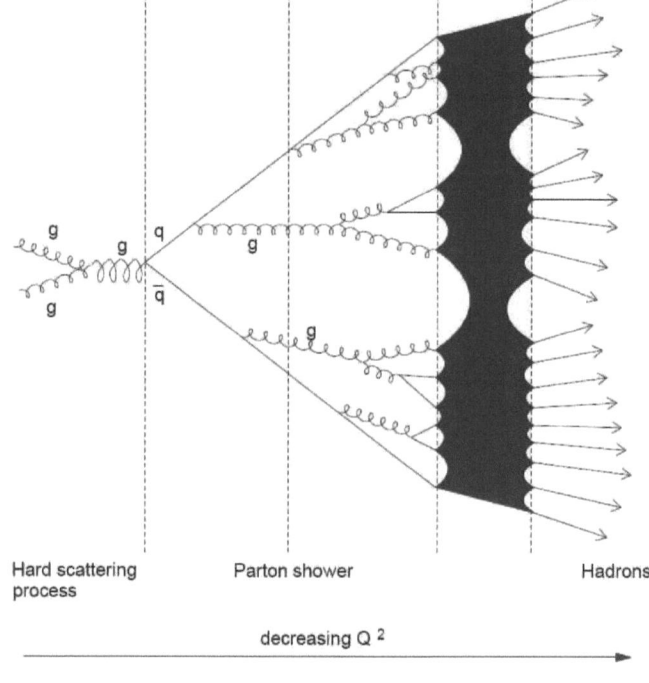

Figure 2.4: Hadronization of color-charged partons into color-neutral hadrons [6]

- The **model of independent hadronization** is the oldest model describing the hadronization process. Here, every quark hadronizes for its own with randomly chosen quark-antiquark-pairs of the vacuum. According to a probability function, the hadron gets a certain fraction of the available energy and momentum.

- A preferable model is the **cluster model**. After the parton shower, all gluons split into $q\bar{q}$- or diquark-antidiquark-pairs. Neighboring quark-antiquark-pairs resulting from such a splitting can build a color singlet cluster due to color-interaction (see figure 2.5). These clusters finally

---
[12]This process is called **hadronization** and displayed in Figure 2.4. As a long-distance process, only small momenta are transfered during the hadronization. This is why the flow of quantum numbers as well as the transfer of energy of the hadrons have their origin mainly from parton-level [13]. This relation is called **local parton-hadron duality** [14].

2.3. Hadronization

decay into hadrons (see [13] for more details). The cluster model is used by the Monte Carlo event generator HERWIG (see chapter 5.3).

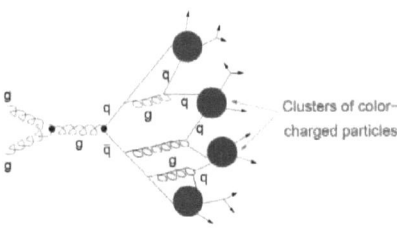

Figure 2.5: Illustration of the cluster model [6]

- A widely used hadronization model, being also implemented in the Monte Carlo event generator PYTHIA (see chapter 5.2.1), is the **(Lund) string model**. After the hard interaction, the color field lines between the partons can be found in color drift tubes. These tubes behave like strings with a constant tension $k \approx 1$ GeV/fm. If the distance between the partons increases, the potential energy rises until enough energy is gathered to build a hadron. Then, the string breaks and forms a $q\bar{q}$-pair. At this stage, the system consists of two color singlets. If one of them has again enough energy available, the described process is repeated. When emitting a gluon, the string is stretched over the gluon, appearing as a "bend" (with momentum and energy) inside the string (see figure 2.6) [6, 11, 12, 15].

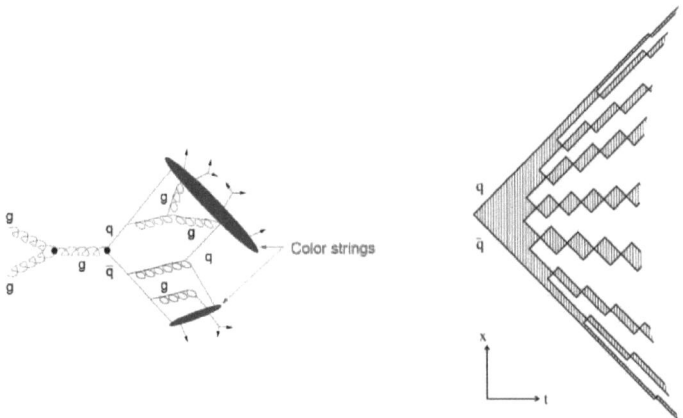

Figure 2.6: Illustration of the string model [6], [16]

2.4 Determination of α_s

In order to determine α_s, processes are needed where gluons take part, as gluons couple with the strength α_s to colored particles. In figure 2.7 (left) an e^+e^- collision with three jets[13] in the final state is shown. One of the jets has its origin in the emission of a gluon. The cross section is in this case proportional to α_s. At proton proton collisions this is slightly different, because a gluon can be exchanged between two final state particles (see figure 2.7, right). As both particles couple with the strength α_s to the gluon, the cross section is proportional to α_s^2.

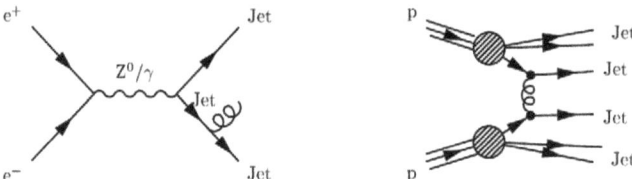

Figure 2.7: Left: e^+e^- collision where a real gluon is emitted in the final state. The cross section is proportional to α_s. Right: pp collision where a gluon is exchanged. The cross section is in this case proportional to α_s^2 [2].

Due to their direct proportionality to α_s, jet-rates provide a good possibility to determine the strong coupling constant.
The exclusive 3-jet-rate

$$R_3 = \frac{\sigma_{3Jets}}{\sigma_{2Jets} + \sigma_{3Jets}} \tag{2.6}$$

is at leading order (LO) proportional to α_s.
At next-to-leading order (NLO), the 3-jet-rate becomes

$$R_3 = A + B , \tag{2.7}$$

where A stands for a LO term (being proportional to α_s) and B for a term at NLO (being proportional to α_s^2).

The next-to-leading order calculation has to deal with unresolved partons, resulting in collinear and infrared divergences. The measurement of jet-rates is hence only possible in certain areas of the phase-space (see figure 2.8).
This problem can be solved by using infrared and collinear safe observables, like the d_{23} flip-values of the exclusive k_T jet algorithm[14].
Jet-rates as inclusive measurements depend strongly on details of the final state, like the hadronization (see chapter 2.3), the parton density function (see chapter 2.5), the Underlying Event (see chapter 2.6.2) or the jet-energy-scale[15] (see chapter 4.5.2). This leads to a huge systematic uncertainty.
Moreover, the entries of the jet-rates are correlated to each other. Therefore, the uncorrelated, differential jet-rates have been studied in this analysis. They have also the advantage that many uncertainties almost cancel out.
In order to study the differential jet-rates, the jet multiplicity has been forced to 3 in this anlysis. The

---

[13] Jets are objects consisting of particles after the haronization, which are close together (depending on the jet algorithm either geometrically or in momentum space). For more details on jets, see chapter 4.

[14] The value d_{23} describes the transition from 3 to 2 reconstructed jets (see chapter 4.3.2).

[15] The jet-energy-scale calibrates the energy measurement of a calorimeter detector to the true energy of a particle jet (or parton jet).

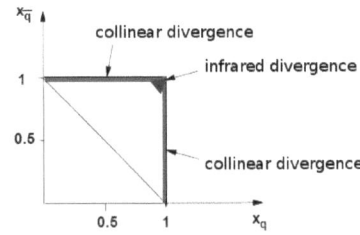
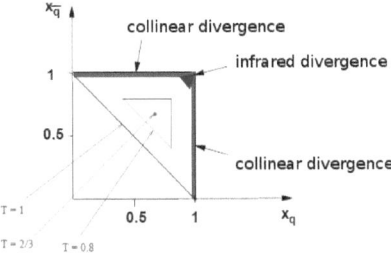

Figure 2.8: Collinear and infrared divergences. Variables like thrust ($T = max_{\hat{n}_T} \frac{\Sigma_i |\vec{p}_i \cdot \hat{n}_T|}{\Sigma_i |\vec{p}_i|}$) (bottom) exclude divergent areas in the phase-space. x stands for the fraction of the proton's momentum [2].

cross section of 3-jet-events is in LO proportional to α_s^3 and to α_s^4 in NLO.
With $R_2 = 1 - R_3 - R_4$ the differential 2-jet-rate becomes (if the 4-jet-rate is neglected[16])

$$D_{23} = \frac{\Delta R_2}{\Delta d_{23}} = -\frac{\Delta R_3}{\Delta d_{23}} = \frac{\Delta A(d_{23})}{\Delta d_{23}} + \frac{\Delta B(d_{23})}{\Delta d_{23}} = \frac{1}{N} \times \frac{\Delta N}{\Delta d_{23}} . \qquad (2.8)$$

It has been pointed out that α_s is not a constant, but changes its value depending on Q. Taking this dependency into account, the above formula becomes

$$D_{23}(Q) = \frac{\Delta A(d_{23}, Q)}{\Delta d_{23}} + \frac{\Delta B(d_{23}, Q)}{\Delta d_{23}} = \frac{1}{N(Q)} \times \frac{\Delta N(Q)}{\Delta d_{23}} , \qquad (2.9)$$

with $\frac{\Delta A(d_{23},Q)}{\Delta d_{23}}$ depending on $\alpha_s^3(Q)$ and $\frac{\Delta B(d_{23},Q)}{\Delta d_{23}}$ depending on $\alpha_s^4(Q)$.
To determine the strong coupling constant, the LO term $\frac{\Delta A(d_{23},Q)}{\Delta d_{23}}$ and the NLO term $\frac{\Delta B(d_{23},Q)}{\Delta d_{23}}$ are both taken from calculations with NLOJET++ [17]. $\frac{1}{N(Q)} \times \frac{\Delta N(Q)}{\Delta d_{23}}$ represents the real data. The real data are such described by a LO and a NLO term, each having different α_s dependencies. Applying fits between real data and the calculated LO and NLO terms (provided by NLOJET++) then yield α_s.

---

[16] R_4 has been neglected, as its NLO calculation is not implemented in the program NLOJET++ (see chapter 5.1) used for the determination of α_s. The measurement has therefore been done in regions where the fraction of the 4-jet-rate is small (see chapter 6.3.3).

2.5 Parton Distribution Functions

As already mentioned, the proton is not an elementary particle. Besides the three valence quarks (uud), being bound together via gluons, it contains sea quarks and gluons (see figure 2.9).

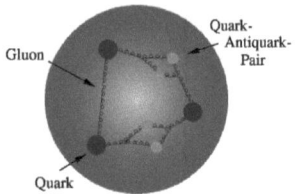

Figure 2.9: Buildup of a proton [18]

Sea quarks are virtual quark-antiquark-pairs, which effective quantum numbers are annihilated on average. They appear at scattering processes because of their electrical charge [3].
The structure of the proton is described by structure functions (see figure 2.10).

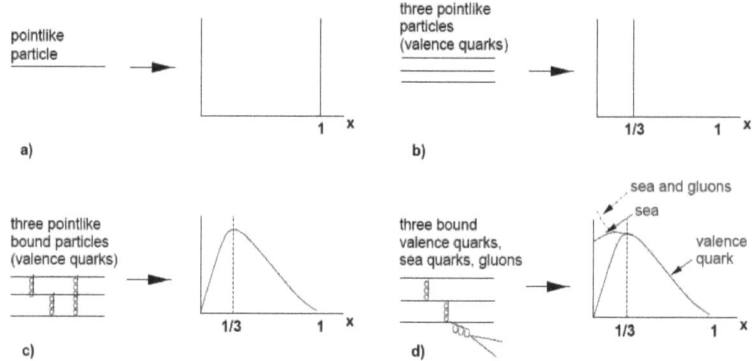

Figure 2.10: Structure functions of a) a pointlike particle, b) a particle composed of three pointlike particles, c) a particle consisting of three bound quarks and d) a proton composed of three valence quarks, sea quarks and gluons [19]. x stands for the fraction of the proton's momentum.

Based on the longitudinal and transverse polarization the structure function

$$F_2(x) = \sum_i e_i^2 \, x \, f_i(x) \qquad (2.10)$$

is discerned from $F_1(x)$. The two functions follow the Callan-Cross relation $2xF_1(x) = F_2(x)$.
$F_2(x)$ describes the superposition of partons i with charge e_i and momentum fraction x [20].
The **parton distribution function (PDF)** $f_i(x)$ parametrizes the probability that the i-th parton carries a fraction x of the original momentum - the rest is assigned to the proton residual (called beam

2.6. Background Processes

remnant).
Therefore, when a collision takes place, two partons with the according fractions of momentum x_1 and x_2 perform a hard interaction (see figure 2.11) [21].

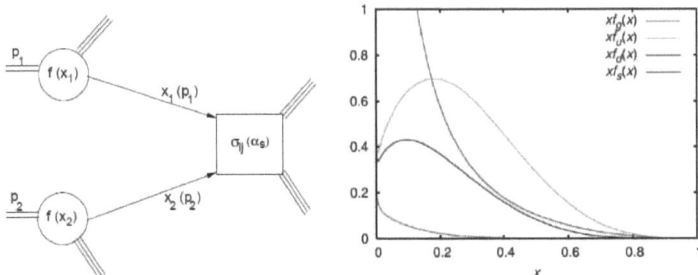

Figure 2.11: Illustration of the factorization of a proton-proton scattering. Left: Parton 1 with the momentum fraction x_1 and the PDF $f(x_1)$ interacts hard ($\sigma_{ij}(\alpha_s)$) with parton 2 (x_2, $f(x_2)$). Right: PDFs (here: CTEQ6) for gluons, up-, down- and strange-quarks at $Q^2 = 2$ GeV versus the longitudinal momentum fraction x [6].

The right plot (using CTEQ6[17]) shows that the gluons dominate in regions where x is small, in contrast to the up- and down-quarks: their fractions of the longitudinal momentum of the proton rise with increasing x.
The PDFs can such be used to calculate the luminosity of the partons in hard collisions. α_s (depending on Q) influences the cross section of the hard scattering process [6].
For more details on PDFs see e.g. [23].

2.6 Background Processes

The complex structure of the protons (see chapter 2.5), the high luminosity of up to $\mathscr{L} = 10^{34}$ cm^{-2}s^{-1} as well as the huge center of mass energy of up to $\sqrt{s} = 14$ TeV involve several problems, being less important at former experiments.
In this chapter processes are described, which take place when bunches of particles collide. The main focus is placed on background processes, overlaying a hard 2 $parton \rightarrow 2$ $parton$ collision and therefore influencing its measurement. At hadron-hadron collisions those soft processes have the largest cross section and are therefore quite important.

2.6.1 Minimum Bias

When two bunches of particles cross each other, the most likely interactions that appear are soft and not hard 2 $parton \rightarrow 2$ $parton$ collisions. Soft means that only a small amount of transverse momentum is transferred. The perturbative Quantum Chromo Dynamic (see chapter 2.2) is very successful at describing hard processes. Unfortunately it cannot be applied when energies become small. Therefore, approximations and models are necessary for these soft interactions [24].
The total cross section predicted at $\sqrt{s} = 7$ TeV is as follows [25] (see figure 2.12):

$$\sigma_{tot}(114.6 \text{ mb}) = \sigma_{el}(24.8 \text{ mb}) + \sigma_{sd}(12.0 \text{ mb}) + \sigma_{dd}(6.2 \text{ mb}) + \sigma_{hc}(71.6 \text{ mb}) \ . \quad (2.11)$$

---
[17]CTEQ stands for the Coordinated Theoretical-Experimental Project on QCD. Several different PDFs (which names are composed of CTEQ and a certain number) have been developed by the CTEQ group [22].

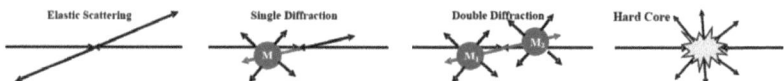

Figure 2.12: Components of the cross section. From left to right: elastic scattering, single diffraction, double diffraction and hard core [26]

σ_{el} stands for the cross section of the elastic scattering, σ_{sd} for the single diffraction, σ_{dd} represents the cross section of the double diffraction and σ_{hc} symbolizes the physically interesting part: the hard core (HC) events. This last component contains soft as well as hard collisions (the hard scattering, i.e. the hard component of the HC, is described in chapter 2.6.2).

The single diffraction can be imagined as the diffraction of the matter wave of one proton at the "disk" of the other proton. The resulting hadrons do not have any color connections to the protons or to the partons of the protons. If the described process is also true for the second proton, it is called double diffraction.
More interesting is the soft component of hard core events (see figure 2.13), being also called **Minimum Bias (MB)**.
At each bunch crossing the design luminosity of LHC will lead to an average of about 23 of these

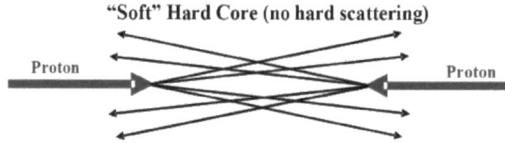

Figure 2.13: Soft hard core component [26]

inelastic, soft events (at the Tevatron[18] only 4 of these inelastic events appeared on an average). The first run periods of LHC had to deal with a maximum average of 3.78 events per bunch crossing [28]. Only a small amount of transverse momentum is transferred at soft events and the direction of the outgoing partons is just slightly different compared to the original hadrons. If a hard scattering process appears at a bunch crossing, the outgoing particles are overlaid by those soft contributions coming from interactions of protons not taking part in the hard scattering process.
It is common to define Minimum Bias as non-diffractive, inelastic interactions [29]. However, there is no consistent definition. Finally, it depends on the used trigger[19] what is considered to be MB in an event [26].

2.6.2 Underlying Event

In order to find interesting and potentially new physics, processes are needed where large transverse momenta are transferred. These events are called hard scattering (see figure 2.14).
Unfortunately, additional soft contributions - commonly known as **Underlying Event** (see figure 2.15, right) - occur at a hard 2 *parton* → 2 *parton* scattering event (independently of the luminosity).

---

[18]The Tevatron is a proton-antiproton accelerator at the Fermilab near Chicago with a luminosity of about 2×10^{32} cm^{-2}s^{-1} and a center-of-mass energy of 1.96 TeV [27].

[19]A trigger is an event-filter. The trigger system of the ATLAS detector is described in chapter 3.2.6.

2.6. Background Processes

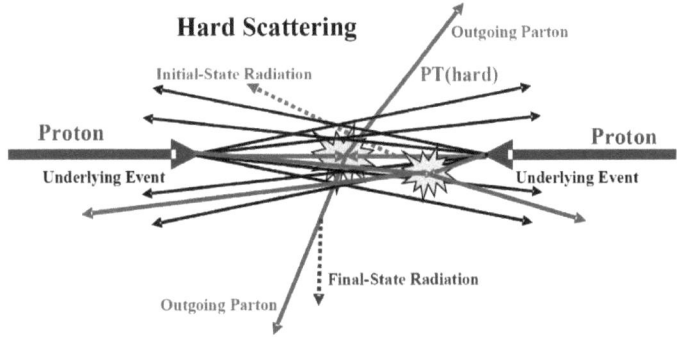

Figure 2.14: Components of the hard scattering with initial (ISR) and final state radiation (FSR) as well as contributions of the UE [26]

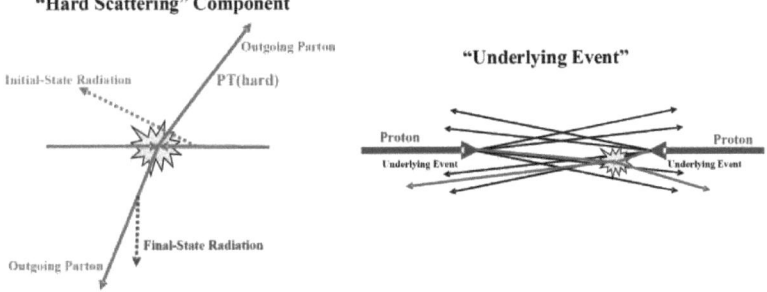

Figure 2.15: Components of the hard scattering, divided in the hard component (left) an the contributions of the UE (right) [26]

The hard scattering consists of particles resulting from the hadronization (see chapter 2.3) of the two outgoing partons. Initial state radiation (ISR) and final state radiation (FSR) (see figure 2.16), i.e. the emission of gluons (or quarks) before or after a collision are commonly assigned to the hard process (however, it should be mentioned that some authors allocate ISR to the UE, because they are experimentally difficult to separate from UE). The energies of the original protons are diminished by the fraction of the ISR and therefore also the available energy for the hard scattering process is reduced.

The term Underlying Event stands for everything except the hard scattering process. It contains **beam remnants** as well as particles resulting from soft or semi-soft **multiple (parton) interactions (MPI)** (see figure 2.17) [26].

Beam remnants are all partons not actively taking part in the hard interaction. If e.g. a down-quark scatters, the remaining up-quarks build (together with other particles) the beam remnant. As these particles are color charged and the proton neutral, they are color connected with the hard interaction

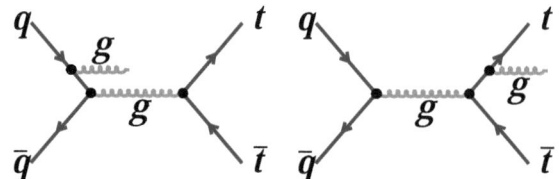

Figure 2.16: Initial (left) and final state radiation (right) [30]

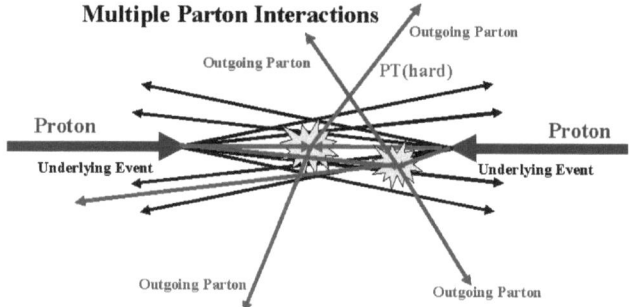

Figure 2.17: Multiple parton interactions [26]

and they are therefore a part of the fragmentation system [21].

In addition there is the possibility that partons not participating in the hard scattering interact inelastically among each other or with partons of the hard scattering proton. The interactions between several partons inside one proton is called multiple parton interaction. MPIs are almost always soft, resulting in particles with mostly small transverse momenta with respect to the beam direction.

The UE depends on the hard scattering, because it has the same primary vertex and it is in addition energy, color and flavor correlated. It is also not equal to the Minimum Bias, although it has a similar phenomenology (however, for some authors the MB is a pars pro toto of the Underlying Event).

In a single event, it is not possible to palpably determine the origin of a particle. No matter which observable is examined, it will always contain fractions of the hard scattering and of the Underlying Event [31].

2.6.3 Pile-up

Minimum Bias and Underlying Event together are called **Pile-up** or **Event-Pile-up** (however, it is also common to use the term synonymously for Minimum Bias).

Furthermore, the term Pile-up is used for the **Detector-Pile-up**, designating the overlay of several events due to the slow read out speed. In the liquid argon (LAr) calorimeter of ATLAS (see chapter 3.2.4) the electronic pulse has a duration of about 600 ns. At the design performance of LHC, a bunch crossing will appear every 25 ns. Hence, an interesting event is likely to be overlaid by particles coming from another bunch crossing [32].

Chapter 3

LHC and ATLAS

This chapter is divided into three parts: the circular particle accelerator LHC, the ATLAS detector and the data processing via the data & computing grid. In this collider experiment, protons scatter with other protons, leading to a very high achievable center-of-mass energy. Protons have the advantage that they do not suffer significantly from synchrotron radiation due to the mass dependency of m^{-4} - in contrary to electrons, being used for the predecessor-experiment LEP. Consequently, the LHC holds the world record for having the highest collision energies, although it has been run with only half of its design center-of-mass energy of 14 TeV until now.
This chapter is based on [4–6, 11] with the parameters mainly taken from [33–35].

3.1 LHC

One of the largest physics experiment ever built on earth is the Large Hadron Collider (LHC) near Geneva with a circumference of 27 km (see figure 3.1). This circular particle accelerator was built in the existing tunnel of the precursor experiment LEP, 100 m under swiss and french territory. Before the protons are injected into the two oppositely running vacuum pipes of the main accelerator, they are brought to 450 GeV by various pre-accelerators, like PSB[1], PS[2] and SPS[3]. At LHC, the protons are gathered in thin bunches of 10^{11} particles. They are finally accelerated to 3.5 TeV by running several times through the same accelerator cavities. Up to now, a maximum of 348 colliding bunches inside LHC has been reached [28]. Superconducting bending magnets are cooled to about 2 K by suprafluid helium and guide the particles, which are accelerated almost to the speed of light until they collide at a center-of-mass energy of 7 TeV (being upgraded to 14 TeV in 2014/15)[4]. The design luminosity of LHC is $\mathscr{L} = 10^{34}$ cm^{-2}s^{-1}. The accelerator is optimized to have a bunch crossing every 25 ns, corresponding to a clock rate of 40 MHz.
In order to detect the particles after a collision took place, four independent detectors have been installed at the intersection points: ATLAS (see chapter 3.2), CMS[5], ALICE[6] and LHCb[7]. ATLAS and CMS are universal detectors and are therefore sensitive to a broad range of physical phenomena.

---

[1] Proton Synchrotron Booster
[2] Proton Synchrotron
[3] Super Proton Synchrotron
[4] Besides the collision between protons, also heavy Ions (Pb-Pb) are brought to collision at a center-of-mass energy of $\sqrt{s} = 5.52$ TeV per nucleon pair. The luminosity is designed to reach up to 10^{27} cm^{-2}s^{-1}.
[5] Compact Muon Solenoid
[6] A Large Ion Collider Experiment
[7] Large Hadron Collider beauty experiment

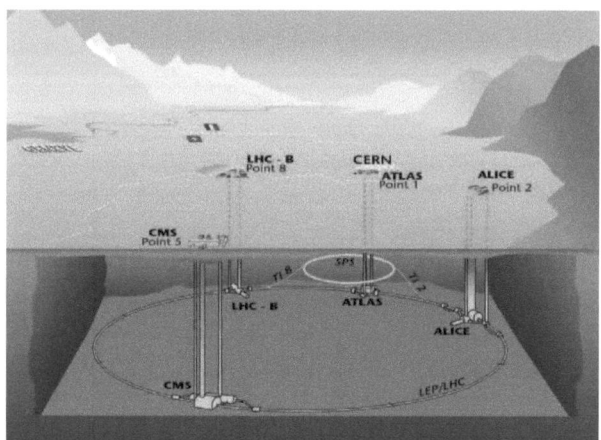

Figure 3.1: The LHC accelerator at CERN [36]

3.2 ATLAS

With a length of 43 m, a diameter of 25 m and a weight of about 7000 t, the ATLAS detector is the largest detector at the LHC. The name ATLAS has formerly been an acronym for **A T**oroidal **L**HC **A**paratu**S** and is used nowadays as a proper name, referring to Atlas from the Greek mythology, who was doomed by Zeus to carry the sky on his shoulders (see picture 3.2) [7].

Figure 3.2: Atlas sculpture in front of the Rockefeller Center in New York. A stylized drawing of this statue is used as the logo for the ATLAS experiment [37].

The ATLAS detector is constructed in several layers, where each layer is sensitive to different particles. In the middle, there is the inner track detector, being surrounded by a solenoid magnet. Then follow the electromagnetic and the hadronic calorimeters and finally the muon system, being

3.2. ATLAS

located inside a toroidal air core magnet. Table 3.1 shows which particles are typically detected in which part of the detector.

	ID	EC	HC	MS
Electron	x	x		
Muon	x		x	x
Charged Hadron	x		x	
Neutral Hadron			x	
Photon		x		
Neutrino				

Table 3.1: Detection of particles in the inner detector (ID), the electromagnetic calorimeter (EC), the hadronic calorimeter (HC) and the muon spectrometer (MS)

The subsystems are divided into a barrel and two end cap regions. Figure 3.3 illustrates the ATLAS detector.

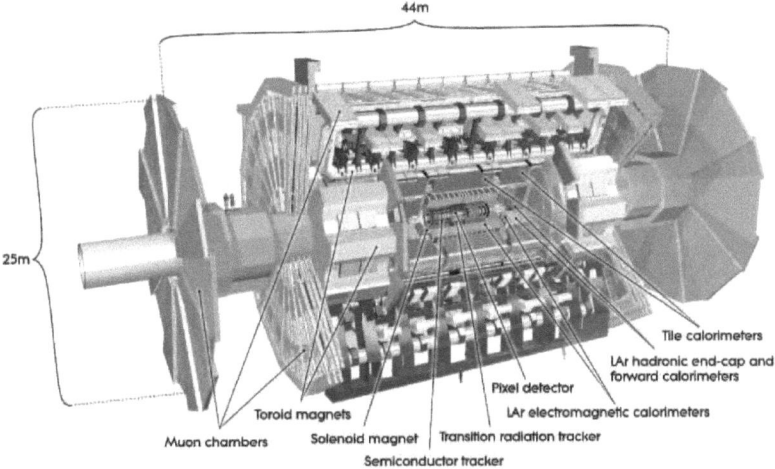

Figure 3.3: The ATLAS detector [38]

3.2.1 Coordinate System

In the right handed coordinate system of ATLAS (see figure 3.4), the z-axis runs along the beam axis. The x-axis points from the collision point to the middle of the LHC accelerator ring, the y-axis upwards.

The **azimuthal angle** Φ is measured perpendicularly to the beam axis, where $\Phi = 0$ is equal to the positive x-axis.

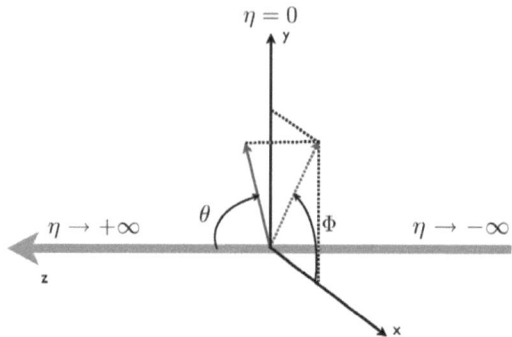

Figure 3.4: The coordinate system at ATLAS [39]

The **polar angle** θ is measured coming from the positive z-axis ($\tan\theta = \frac{r}{z}$, with $r = \sqrt{x^2 + y^2}$). Instead of the polar angle the **pseudorapidity** is usually used, having the advantage that the difference of two η-values is invariant under Lorentz boosts along the z-direction. The pseudorapidity is defined as follows:

$$\eta = -\ln(\tan\frac{\theta}{2}) \ . \tag{3.1}$$

Another important value is the **transverse momentum** p_T, representing the projection of a particle's momentum to the plane perpendicular to the beam axis.

3.2.2 Magnet System

As the supraconducting magnet system does not detect any particles itself, but helps other detector components with this duty, its description is set in front of the detector subsystems.
The magnet system consists of a central solenoid (CS) and three toroid magnets.

- The **central solenoid** is 5.3 m long and has a radius of 1.2 m. It surrounds the inner detector (see chapter 3.2.3) and generates a magnetic force of 2 T with a maximum of 2.6 T. In this magnetic field, the tracks of charged particles are curved in the xy-plane.

- The **toroid magnets** generate the magnetic field for the muon spectrometer (see chapter 3.2.5). The magnet system consists of eight supra conducting, toroidal air coils, being cooled to 4.5 K by liquid helium. In the end caps two additional magnets are installed. Their fields are overlapping the fields of the toroid magnets. The toroidal magnetic field has an average of 0.6 T.

From inside out the ATLAS detector consists of the following subsystems:

3.2.3 Inner Detector

The **inner detector** with a length of 6.2 m and a diameter of 2.1 m is located around the interaction point (see figure 3.5). In this part of the detector, the particle tracks, which are bend by the adjacent solenoid magnets, are measured and the momenta of the charged particles determined.

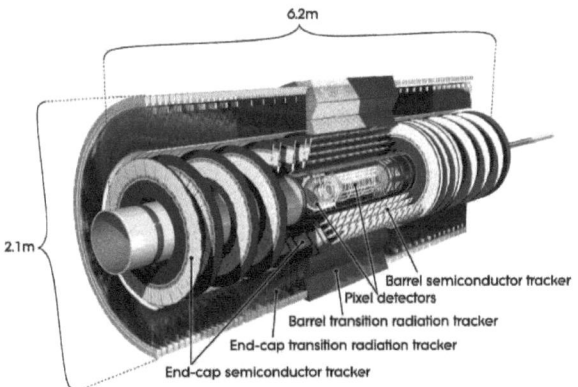

Figure 3.5: The inner detector [38]

The inner detector also has a substructure and consists of three parts:

- 1500 cylindrical and 700 disk shaped semiconductor modules build up the **pixel detector**. Each pixel module has an area of 13.35 cm^2 with 61,440 pixels. The silicon pixel counters are arranged cylindrically in three layers around the beam axis. In addition, five slices are installed on each end, so that almost the whole solid angle is covered. The pixel detector provides three measuring points per particle track, being used to reconstruct the vertices. The resolution is 12 μm in the $R\Phi$- and 66 μm in the z-direction. The ability to measure short living particles and the resolution of the impact parameter are mainly set by this part of the detector.

- The **semiconductor tracker** (SCT) is composed of eight layers of silicon strip detectors, allowing precision measurements of up to eight additional points of the particle tracks in the $R\Phi$- (resolution: 16 μm) and z-area (resolution: 580 μm). The SCT contributes to the measurement of the momenta, the impact parameter and the vertex positions. It covers an area of $|\eta| < 2.5$.

- Finally, the **transition radiation tracker** (TRT) completes the inner detector. The particle tracks are measured like in drift chambers. The electrons are detected via additional transition radiation in pipes, filled with xenon. In this way, additional 36 track points with a resolution of 0.170 mm for charged particle tracks with $p_T > 0.5$ GeV at $|\eta| < 2.5$ are gathered. A good separation of electrons from pions is therefore possible with the TRT.

3.2.4 Calorimeter

The energies of the particles are measured in the **calorimeter** via absorption (see figure 3.6). This part of the detector has again a substructure, consisting of an electromagnetic and a hadronic calorimeter. Both are sampling calorimeters, i.e. alternating slices of energy absorbing materials with high density, and gaps, where the resulting particle showers are measured.

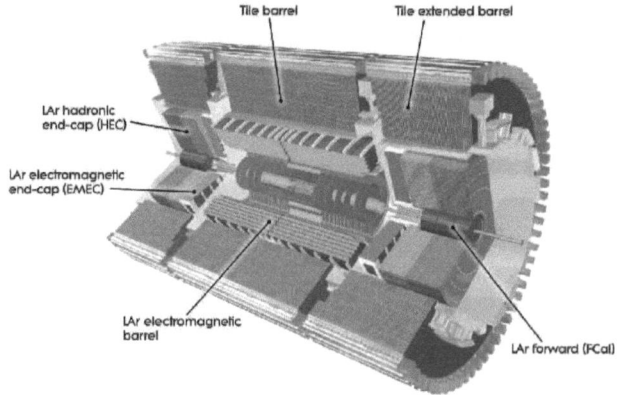

Figure 3.6: The calorimeter [38]

- In the **electromagnetic calorimeter** (EM) the energy of mainly electromagnetically interacting particles (above all electrons, positrons and photons) is absorbed. Muons and hadrons lose a fraction of their energy, but are still able to reach other parts of the detector. After the interaction with lead the resulting particle showers are detected in 2.1 mm thick gaps, filled with liquid argon (LAr). The resolution for electromagnetic showers is $\Delta E/E = 10\%/\sqrt{E/\text{GeV}}$. The EM calorimeter covers a region of $1.4 \leq |\eta| \leq 3.2$ for the end cap and up to $|\eta| = 1.475$ for the barrel region respectively.

- Strongly interacting particles, i.e. hadrons, are absorbed in the **hadronic calorimeter** (HC). While lead absorbers and scintillator plates are used in the barrel region ($|\eta| \leq 1.7$) to detect hadronic showers, copper and wolfram absorbers are utilized in the end caps ($|\eta| \leq 3.2$) and in forward direction as well as liquid argon as sampling material. The accuracy of the HC with a value of $\Delta E/E = 50\%/\sqrt{E/\text{GeV}}$ is significantly lower than the accuracy of the EM.

3.2.5 Muon Spectrometer

The **muon spectrometer** identifies and measures muons. As their ionization is rather small they pass the inner detector and the calorimeter almost undisturbed - in contrary to the other particles - and can therefore clearly be identified (neutrinos on the contrary can only be indirectly detected with the ATLAS detector).
Based on the magnetic deflection of the muon tracks by the toroid magnets, the momenta can be estimated. The precision measurement of the track coordinates in the largest part of the η-region is done by **monitored drift tubes (MDTs)**. MDTs are made up of three cylindrical layers of drift tubes (plus three at the end caps), being filled with a mixture of argon and carbon dioxide.

3.2. ATLAS

At large values of η and in the vicinity of the beam axis, **cathode strip chambers (CSCs)** are used. These are multi-wire proportional chambers with a smaller fragmentation in order to deal with the high particle fluxes.

To trigger on muons, **resistive plate chambers (RPCs)** are used in the central region and **thin gap chambers (TGCs)** in the end caps. The end caps of the muon spectrometer are shown in figure 3.7.

Figure 3.7: End caps of the muon spectrometer [40]

3.2.6 Trigger

Due to the high collision-rate, the ATLAS detector accumulates one terabyte of data every second. As it is not possible to store all of these data, the bunch crossing rate of 40 MHz has to be reduced to an event-rate of about 200 Hz for permanent storage. In order to distinguish interesting from not interesting (e.g. low energetic background) events, an efficient trigger system is needed.

The ATLAS trigger is composed of three parts:

- The first reduction is done by the hardware trigger **level one (LVL1)**, selecting events with the help of signals from the calorimeter and the muon spectrometer. Its aim is to identify the bunch crossing, where an interesting event took place and to mark **regions of interest** (RoI), i.e. areas in the detector, contributing interesting data to the event.
  During the latency of 2.5 μs all data is stored in pipeline storages: uninteresting events are removed, and events passing the trigger criteria are stored in the **readout buffer** (ROB). In this way, the event-rate is reduced to 75 kHz.

- Then, every RoI is analyzed again. Based on selection algorithms implemented in software, the **level two (LVL2)** trigger makes further reductions, having access to the full resolution of the RoIs as well as the whole inner detector. After the LVL2 trigger, the event-rate is reduced to **1 kHz**.

- The final decision over an event is made by the (software based) **event filter** (EF) on a computer cluster. The EF also classifies and saves the remaining data according to their event types. The final acceptance rate of 200 Hz makes it possible to store the selected events permanently for physics analysis. Those events are distributed to world wide computer farms (see chapter 3.3).

The figure 3.8 gives an overview of the trigger system.

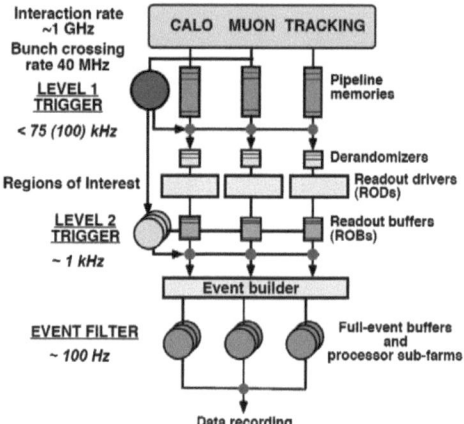

Figure 3.8: The trigger system of ATLAS [41]

In this thesis, the triggers L1_J15 and L1_J30 have been applied, as only information from LVL1 has been used to select events in the first run periods. These level one triggers consider jet elements, which are towers of 0.2×0.2 in the $\eta \times \phi$ space of the electromagnetic and hadronic calorimeters. If the transverse energy of the cluster has a local maximum within a region $\Delta\eta \times \Delta\phi = 0.4 \times 0.4$ (see figure 3.9) and $|\eta^{jet}| < 3.2$, the jet is reconstructed at LVL1. It passes the trigger if the transverse energy-deposition inside $\Delta\eta \times \Delta\phi = 0.8 \times 0.8$ (4 × 4 jet elements) is above a certain threshold (in this case 15 GeV and 30 GeV respectively) [42].

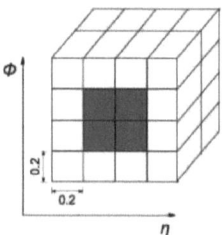

Figure 3.9: The trigger tower of size 4 × 4 jet elements ($\Delta\eta \times \Delta\phi = 0.8 \times 0.8$) inside the electromagnetic and hadronic calorimeter. In grey the cluster transverse energy with a maximum inside a region of 2 × 2 jet elements ($\Delta\eta \times \Delta\phi = 0.4 \times 0.4$) is illustrated [42].

3.3 Data & Computing Grid

In order to cope with the design trigger rate of 200 Hz (relating to 1 PB of data per year), sophisticated analysis software as well as a new, powerful computing infrastructure are needed. Therefore, the **data & computing grid** was invented: a world wide network of computer clusters, where offline reconstruction of observables of the recorded events and the user analyses are done. Computing centers all over the world are connected in a hierarchically order (see figure 3.10). Starting from Tier-0 at CERN, the data is processed for the first time and distributed world wide to the adjacent centers, the Tier-1, where the data processing is pursued, stored and distributed to the hierarchically next centers. These are called Tier-2, which perform larger physics analyses, Monte Carlo simulation sample productions (see chapter 5) and also store selected data sets. Finally, the Tier-3 clusters are used for smaller user analyses and test jobs.

The full datasets should solely be available at Tier-0, while the other Tier centers should only keep fractions to distribute the load. Instead of downloading the datasets needed for a user analysis, the procedural method is to send the analysis to the data on the data & computing grid, let it process there and finally get the results back to the local computer [5].

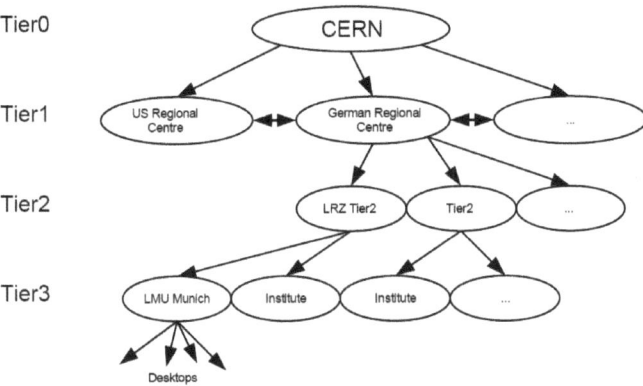

Figure 3.10: The structure of the data & computing grid [5]

Chapter 4

Jets

This thesis focuses on the determination of α_s via the differential 2-jet-rate. Therefore, jets are the most important signature for these studies.
In this chapter, the production and reconstruction of jets is described. As there are several jet algorithms, assigning particles to a jet in different ways, the most common clustering methods are shown. These are the cone, the k_T and the anti-k_T algorithms.

4.1 Jet Production

Protons are not elementary particles, but have a substructure. As already mentioned in chapter 2.5, they are composed of quarks and gluons. In an ideal case the collision of two protons leads to a high energetic interaction between a parton of one proton with a parton of the other proton (in reality, there can appear further interactions, see chapter 2.6.2). The two scattered partons appear at large angle (respective to the beam axis) and emit gluons and quarks. These quarks and gluons radiate gluons themselves, which decay into quark-antiquark-pairs - a parton shower is induced. The bunches of these high energetic partons are called parton jets. These color-charged particles hadronize to color-neutral particles - the particle jet is formed (see figure 4.1). After the hadronization, the jets consist of stable and long-living particles, such as pions. The particles are absorbed in the calorimeter, clustered to jets and assigned to the original parton. For the allocation of the particles and energy depositions in the calorimeter, jet algorithms are necessary [2]. First of all, the cone algorithm will be introduced, followed by the k_T algorithm. Finally, the new standard of the ATLAS experiment, the anti-k_T algorithm, is presented.

4.2 Cone Algorithm

The cone algorithm has been the standard jet reconstruction algorithm at former hadron collider experiments like Tevatron. As it is still quite often used at LHC, it is described in this chapter. However in the last few years it has been replaced more and more by the anti-k_T algorithm (see chapter 4.3.3).
The cone jet algorithm clusters particles inside a fixed cone in azimuthal angle ϕ and pseudorapidity η. The simplest version of the cone algorithm assigns particles to a jet, which are inside a certain cone with radius $R = \sqrt{\Delta \eta^2 + \Delta \phi^2}$ (typically 0.4 or 0.7) around a seed[1] (where $\Delta \eta$ and $\Delta \phi$ represent the differences between the η and ϕ values between the seed and the investigated particles). If a particle is inside the cone, the centroid of this new cluster is recalculated and a new jet axis is defined. Particles outside the cone are not allocated to the jet. This leads to round jets (see figure 4.2, left).

---

[1]Seeds are particles or preclustered objects with a certain minimal transverse momentum p_T (typically a few GeV).

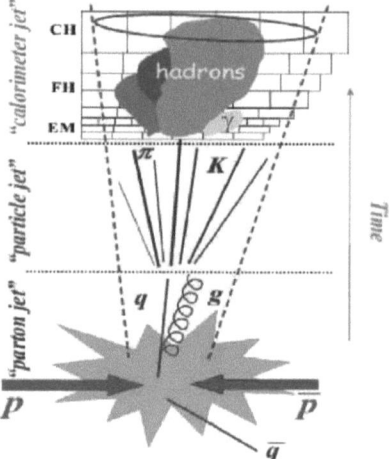

Figure 4.1: The formation of jets. At the beginning, the parton jet is build. The subsequent hadronization leads to the particle jet. Finally, the jets are reconstructed in the calorimeter in accordance with their energy depositions [43].

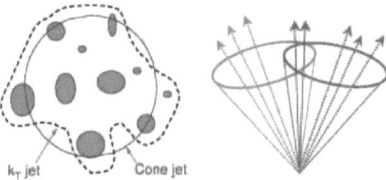

Figure 4.2: Left: Comparison between cone and k_T jets [43]. Right: 2 cone jets overlap each other [43]

Due to the geometrical association, the algorithm has to deal with some difficulties. As the whole detector has to be covered with cones, overlapping jets can appear (see figure 4.2, right). If a particle is for example inside the cones of two jets, further iterations are necessary to assign the particle to one of them. In addition, the cone algorithm is not infrared safe: If a gluon with low transverse momentum is e.g. emitted between two jets, the two jets may be incorrectly merged to one, changing the jet multiplicity in the final state (see figure 4.3).

Furthermore, the cone algorithm is not collinear safe. Collinear means that the angle between a high energetic radiated gluon and the radiating parton is very small. This can also result in a wrong jet multiplicity in the final state, as the cone algorithm cannot cover these areas in the phase-space. Instead of assigning the high energetic gluon to the existing jet, the cone algorithm might find two jets.
However, there are some improved cone algorithms like the seedless, the midpoint, or the SISCone algorithm, solving some of these problems (see e.g. [44]).

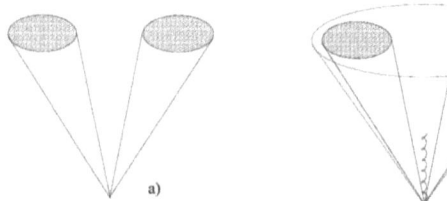

Figure 4.3: The two jets in the final state (left) are merged to one jet (right) because of the emission of a soft gluon. Due to the lack of infrared safety, the jet multiplicity changes from 2 to 1. Soft gluon radiation is large because of the infrared divergences of the cross section [6].

4.3 k_T Algorithm

This chapter summarizes the k_T algorithm described in [45] with a focus on the settings used in this analysis. The k_T algorithm has the advantage that there are no overlapping jets, as the jet size is dynamic and therefore every particle is assigned to exactly one jet. Hence, the jets are not round anymore (see figure 4.2, left). Finally, the k_T algorithm is collinear and infrared safe in each order of perturbation theory [46] and has only a small dependency on hadronic corrections (see chapter 7.2).
Originally, there have been two different kinds of the k_T algorithm: the inclusive and the exclusive mode. The difference is the definition of the hard final state jet and the separation from the beam remnants. In both cases the resolution variables d_{kB} (the distance in momentum-space between an object k and the beam jet B, i.e. proton rest) and d_{kl} (the distance in momentum-space between an object k and an object l) are evaluated for all final state objects h_k and pairs h_k and h_l. The definition of the resolution variables can be chosen among different angular definitions, influencing the behavior of the k_T algorithm in the soft and collinear limits.
In the **angular scheme** (typically used in e^+e^- annihilation analyses) the resolution variables are defined as follows:

$$d_{kB} = 2E_k^2(1 - cos(\Theta_{kB})) \quad \text{and} \tag{4.1}$$

$$d_{kl} = 2min(E_k^2, E_l^2)(1 - cos(\Theta_{kl})) \ . \tag{4.2}$$

Another definition is the ΔR **scheme**:

$$d_{kB} = p_{Tk}^2 \quad \text{and} \quad d_{kl} = min(p_{Tk}^2, p_{Tl}^2) \times R_{kl}^2 \quad \text{with} \tag{4.3}$$

$$R_{kl}^2 = (\eta_k - \eta_l)^2 + (\Phi_k - \Phi_l)^2 \ . \tag{4.4}$$

For hadron-hadron collisions this is the most common choice and therefore used in this thesis. As the distance between two objects in the transverse momentum-space[2] is used, no seeds are needed for the k_T algorithm. Furthermore it considers the characteristic that the decay products have the tendency of having similar momenta.

An alternative definition of R_{kl}^2 is provided by the **QCD emission scheme** (see [45] for further details).

---
[2]The transverse momenta of the particles is in this case defined respective to the direction of the parton (represented by the jet).

Besides the jet resolution variables, the recombination schemes, i.e. how two objects h_k and h_l are merged into a single object with 4-momentum p_{kl}, can be controlled by the user:
The **E scheme** makes a simple 4-vector addition

$$p_{kl} = p_k + p_l, \qquad (4.5)$$

resulting in massive final state jets. The E scheme is the default in FastJet (see chapter 5.1) and it has been used in this thesis - except chapter 7.4, where the E_T **scheme** has been applied.
The E_T scheme is defined via

$$E_{T,ij} = E_{Ti} + E_{Tj}, \qquad (4.6)$$

$$\eta_{ij} = \frac{E_{Ti}\eta_i + E_{Tj}\eta_j}{E_{T,ij}} \quad \text{and} \qquad (4.7)$$

$$\Phi_{ij} = \frac{E_{Ti}\Phi_i + E_{Tj}\Phi_j}{E_{T,ij}}. \qquad (4.8)$$

Although this scheme deals with massless as well as massive input objects, the combined output objects are massless.
Other choices of the recombination scheme are the p_T **scheme**, the p_T^2 **scheme** and the E_T^2 **scheme** (see [45] for further information).

4.3.1 Inclusive Mode

When using the inclusive mode of the k_T algorithm, the distance in momentum-space between a particle and the beam d_{kB} is scaled by the dimensionless parameter R^2 (usually set to 1.0): $d_k = d_{kB} \times R^2$. Due to this scaling, which defines the extent of the jets, the inclusive k_T algorithm behaves similarly to a cone algorithm. In the next step, the smallest distance among all d_k and d_{kl} is found. On the one hand, if d_{kl} is smaller than d_k, the objects h_k and h_l are merged to a new object with momentum p_{kl}.
On the other hand, if a d_k is smaller than d_{kl}, the object k is defined as a jet and therefore removed from the list of objects to be merged. This procedure is repeated until all particles are assigned to jets.
In contrary to the exclusive mode (see chapter 4.3.2) there is no cut-off parameter as a stopping condition. The only parameter influencing the size and the number of jets is R. The low-p_T scattering fragments are therefore not strictly separated from the hard subprocess, meaning that parts of the proton remnants are possibly included in the reconstructed jets. As a consequence, the inclusive k_T algorithm finds a large number of jets.

4.3.2 Exclusive Mode

In this analyis, the k_T algorithm has been applied in the exclusive mode, as it provides flip-values from $n+1$ to n reconstructed jets. These flip-values have been used to investigate the differential jet-rates.
With this algorithm the hard final state is explicitly separated from the soft beam remnants. The stopping parameter d_{cut} (with the dimension of energy squared) defines the hard scale of the process: $\Lambda_{QCD}^2 \ll d_{cut} \leq s$, with s being the squared center-of-mass energy and Λ_{QCD} the only free parameter of QCD with a value of a few hundred MeV.
The flow chart 4.4 visualizes the reconstruction procedure.

First of all, the algorithm searches for the smallest value among all d_{kl} and d_{kB}. This value is called d_{min}. If d_{kB} has the smallest value, the object k is included to the beam jet (i.e. proton rest) and

4.3. k_T Algorithm

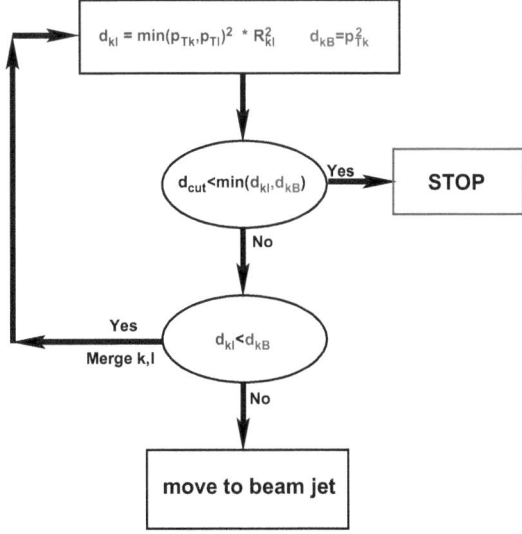

Figure 4.4: Flow chart: jet reconstruction with the k_T algorithm in the exclusive mode

removed from the list. In the opposite case ($d_{kl} < d_{kB}$), the objects h_k and h_l are merged (analogously to the inclusive mode) to a single object.

This merging process is repeated until $d_{min} > d_{cut}$. Then, all remaining objects are classified as jets and the algorithm stops. Hence the d_{cut} parameter defines the maximal distance in momentum-space between two particles. A small value of d_{cut} ($d_{cut} \to \Lambda_{QCD}^2$) leads to many jets in the final state, whereas for a large value ($d_{cut} \to s$) a small jet multiplicity is obtained [6, 45]. (A stopping parameter of 400 GeV^2 e.g. corresponds approximately to a minimum jet momentum of 20 GeV.)

Instead of setting a d_{cut} value, it is also possible to fix the jet multiplicity to a certain value and retrieve the according value of d_{min}. By choosing to stop the merging when 3 jets are reached, it is possible to get the flip-value, where the multiplicity drops from 3 to 2 jets (in the following called d_{23}) - which has been used for this analysis (see figures 4.5).

4.3.3 Anti-k_T Algorithm

The default algorithm used in ATHENA[3] is the anti-k_T algorithm. This algorithm is infrared and collinear safe and in addition behaves like a perfect cone algorithm. The resolution variables are defined along the lines of regular k_T jets besides having negative exponents:

$$d_{kB} = p_{Tk}^{-2} \quad \text{and} \quad d_{kl} = min(p_{Tk}^{-2}, p_{Tl}^{-2}) \times R_{kl}^2 \quad \text{with} \tag{4.9}$$

$$R_{kl}^2 = (\eta_k - \eta_l)^2 + (\Phi_k - \Phi_l)^2 \ . \tag{4.10}$$

---
[3] ATHENA is a software framework for studies of the ATLAS experiment, see chapter 5.4.

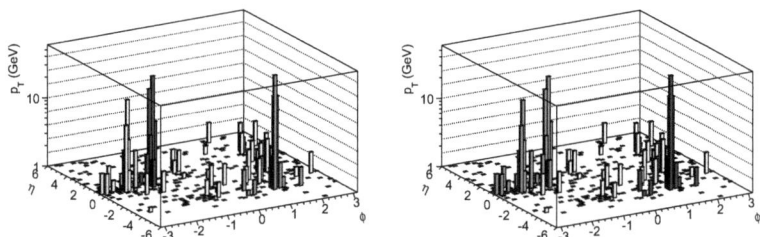

Figure 4.5: Event with 3 jets (left) and 2 jets (right) in the final state [47]. The leading jet is shown in green. At the transition from 3 to 2 jets, the blue and the yellow jet are merged to one jet, having the highest transverse momentum in this event. The value d_{23} describes the transition from 3- to 2-jet-events.

Due to this definition, this algorithm begins with high energetic particles and adds low energetic objects at the end. This is the reason why the anti-k_T algorithm cannot be used for this analysis, because in this case the flip-values of the last clustering steps describe the apposition of soft particles and are therefore not useful for the determination of α_s from gluon radiation at a high energy-scale.

4.4 Inputs to Jet Reconstruction

The above described jet algorithms can be applied on several objetcs, like particles. At collider experiments, particles can only be measured indirectly, e.g. via their energy depositions in the calorimeter cells. In order to combine the 180,000 calorimeter cells of ATLAS (see chapter 3.2) into discrete objects as jet inputs, two generic approaches are available [48]:
On the one hand, there is the **tower grid**, a projective fixed 2-dimensional grid in η and ϕ (grid size: $\eta \times \phi = 0.1 \times 0.1$), which is filled with calorimeter cell energies calibrated at the electromagnetic (em) scale.

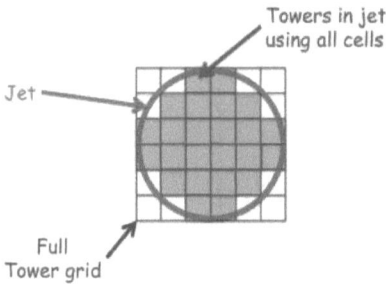

Figure 4.6: Standard calorimeter tower input to jet finding [49]

Noise suppression is performed by using only cells, being selected by a special algorithm [50]. This algorithm is commonly known as the 4-2-0 seed and neighbor noise suppression algorithm. It is based on a signal-to-noise ratio of seed cells and neighbors.

The **topological clustering** (topo-clusters) on the other hand yields 3-dimensional energy clusters. The clusters group calorimeter cells into energy "blobs", representing the energy depositions of particles entering the calorimeter. The results are clusters with mutable numbers of cells. For noise suppression, the 4-2-0 scheme is used.

Instead of considering the calorimeter information, jets can be build from other 4-momentum objects, like **truth particles**[4] or **tracks**. The latter use the track information of the inner detector (see chapter 3.2.3). 3D track jets employ z-clustering in order to ensure that all tracks are coming from the same interaction [48].

4.5 Jet Correction

In order to get rid of detector effects, the calorimeter response and the jet energy have to be adjusted. Finally, events have to be cleaned from jets with bad quality. The details are explained in the following three subsections.

4.5.1 Correcting for Calorimeter Response

The calorimeter (see chapter 3.2.4) determines the energy of jets. However, the calorimeter has to cope with non-linearity of response of the detector to the particles' energy deposits. Therefore, the calorimeter response has to be corrected to ensure proportionality between the energy of the particles and the measured energy.

To calibrate the jets, two different approaches are used: a global and a local calibration. In the **global cell weighting** (or **H1-style**) the correction is done from top to down: First, the jet algorithm runs over uncalibrated (i.e. at em scale) calorimeter towers or topo-clusters to reconstruct the final state objects as jets and MET[5]. Second, cell-by-cell weights are used, depending on the measured cell energy-density and position. This results in jets calibrated to the hadronic scale.

A bottom-up approach is done by the **local hadronic calibration**. In this case, em calibrated topo-clusters are used. In a first step, the calorimeter objects are fully reconstructed and the clusters are calibrated by discriminating the electromagnetic and hadronic clusters. Then, jets and MET are reconstructed from calibrated topo-clusters. The local hadronic calibration also results in jets calibrated to the hadronic scale [51].

As already outlined, the k_T algorithm in the exclusive mode is not the default in ATHENA and jets reconstructed by this algorithm are therefore not stored in the analysis objects. In order to apply this algorithm, the user has to rerun the jet reconstruction, using local calibrated topo-clusters (LCTopo) as input. It is not possible to rerun jets on global weighted clusters, as they depend on the jet algorithm used. Thus, only LCTopo and truth particles (see below) have been used in this analysis.

4.5.2 Jet-Energy-Scale

The above correction of the calorimeter response ensures proportionality between the particle energy and the energy measured in the calorimeter.

In addition, an absolute energy scale is needed, as not every particle created during the scattering

---
[4]Truth particles are all particles having their origin in a quark or gluon. No particles are lost in the reconstruction process as calorimeter effects are neglected. Hence, the particles created by Monte Carlo generators are reconstructed without loss of energy.

[5]MET stands for the missing transverse energy due to undetected particles.

and hadronization process can be detected by the calorimeter, resulting in a diminished jet energy measurement. This is caused e.g. by particles, crossing the calorimeter without losing all of their energy. Moreover, energy can also be lost in non-instrumented materials of the detector in front of the calorimeter, like coils, cables, supports, etc.

These effects (as well as algorithm inefficiencies) are compensated by **jet-energy-scale** corrections. The jet-energy-scale is determined using simulation programs, but shall be derived from measured data later. The software framework ATHENA (see chapter 5.4) can simulate detector effects for Monte Carlo samples and reconstruct jets from calorimeter cells (like it is done with real data). Additionally, it is possible to build jets (from simulated data) with truth particles as inputs. Jets consisting of such particles created by Monte Carlo generators and hitting the calorimeter are commonly known as truth jets. The next step is to compare the reconstructed jet energy with the energy of the truth jets in the according p_T-bins in order to get the jet-energy-scale. Then, it is possible to rescale the measured jet energy in accordance with the energy of the original quark or gluon.

The calibration of the jet-energy-scale in ATLAS is unfortunately only done for the anti-k_T algorithm[6]. In this analysis, a bin-by-bin correction has therefore been applied to correct the exclusive k_T jets for calorimeter effects (see chapter 7.1).

4.5.3 Jet Cleaning

The quality of jets [53] is divided into three groups: bad jets, ugly jets and good jets. If jets are not associated to in-time real energy depositions in the calorimeters, they are called **bad jets**. Possible sources are e.g. hardware problems, LHC beam conditions as well as cosmic ray showers. To identify such jets, there are a number of cuts available (see table 4.1).

	Bad Jet Definition
EM coherent noise	(f_{EM} >0.95 and $|Q|$ > 0.8) or
HEC spike	(f_{HEC} >0.8 and n_{90} ≤5) or
Cosmics - Beam background	$|t|$ > 50 ns

Table 4.1: Definition of bad jets: f_{EM} stands for the electromagnetic fraction, f_{HEC} for the energy fraction in the hadronic end cap calorimeter (HEC), Q for the jet quality (i.e. the fraction of LAr cells with a cell Q-factor larger than 4,000, where Q measures the difference between the measured and the predicted pulse shape that is used to reconstruct the cell energy), t for the jet time, which is computed as the energy squared cells mean time and n_{90} being the minimum number of cells containing at least 90% of the jet energy [53,54]. In chapter 6.3.2 some of these parameters are explained in more detail.

It should be mentioned that newer jet cleaning cuts have been developed (see table 4.2). Unfortunately, they have not been implemented in the ATHENA framework yet. This is the reason why the older cuts from table 4.1 have been used for this analysis. These cuts are only slightly different from the definition of the recent loose cleaning cuts.

With the loose definition, most of the fake jets and missing tails due to detector failures are removed introducing a very small jet inefficiency of $<$ 0.1%. The tight definition leads to very clean data samples with an inefficiency of a few percent.

Ugly jets relate to energy depositions in areas with non-accurate energy measurements, like the transition region between the barrel and the end cap. If jets are neither bad nor ugly, they are called **good jets**. In this thesis, events with one or more bad jets have not been analyzed.

---
[6]The JES at ATLAS currently has an uncertainty of around 5% [52].

4.5. Jet Correction

	Loose	Tight
EM coherent noise	(f_{EM} >0.95 and $|Q|$ > 0.8) or	(f_{EM} >0.90 and $|Q|$ > 0.6) or
HEC spike	(f_{HEC} >0.8 and n_{90} ≤5) or	(f_{HEC} > 1 − $|Q|$) or
	(f_{HEC} >0.5 and $|Q|$ > 0.5) or	(f_{HEC} >0.3 and $|Q|$ > 0.3) or
Cosmics -	$|t|$ > 25 ns or	
Beam background	f_{EM} <0.05 or	f_{EM} <0.10 or
	(f_{max} >0.99 and $|\eta|$ < 2)	(f_{max} >0.95 and $|\eta|$ < 2)

Table 4.2: New definition of bad jets: f_{EM} stands for the electromagnetic fraction, f_{max} for the maximum energy fraction in one calorimeter layer, f_{HEC} for the energy fraction in the hadronic end cap calorimeter (HEC), Q for the jet quality, t for the jet time and n_{90} for the minimum number of cells containing at least 90% of the jet energy [53].

In chapter 6.3.2 some jet cleaning variables are studied and the cuts are applied to real data.

Chapter 5

Analysis Software

The used programs are described in this chapter. First of all, the program NLOJET++ [17] is introduced, allowing the calculation of parton production in leading and next-to-leading order. Unfortunately, there is no hadronization model implemented in NLOJET++. Thus, hadronization effects and the influence of the Underlying Event have been studied with the program PYTHIA [21]. HERWIG [55] has been used to study the systematic uncertainty of the hadronization. Finally, the software framework ATHENA [56] is described. This program has been used to analyze fully simulated data (using Monte Carlos generators[1]) as well as real data.

5.1 NLOJET++

The program NLOJET++ (version 4.1.3) [17] by Zoltan Nagy is a numerical integration program, calculating cross sections (in units of nanobarn) for parton productions. For this analysis, the program has been used to calculate proton proton collisions[2] at a center-of-mass energy of 7 TeV.
NLOJET++ calculates the cross sections of the leading order (LO or born), the next-to-leading order (NLO) contributions, or of both together (full), using the Catani-Seymour dipole subtraction method. This method is modified to make the calculation computationally simpler [57].
The total cross section in NLO accuracy consists of the leading order cross section (i.e. the integration over the fully exclusive born matrix element of k final-state partons in the available phase-space) and the NLO term:

$$\sigma = \sigma^{LO} + \sigma^{NLO} = \int_k d\sigma^B + \sigma^{NLO} \ . \tag{5.1}$$

The NLO contribution is composed of the real correction, being the integral of the born matrix element of $k+1$ final state partons, and the virtual correction. The latter is the integral of the interference term between the one-loop amplitudes of k final state partons and the born-level:

$$\sigma^{NLO} = \int_{k+1} d\sigma^R + \int_k d\sigma^V \ . \tag{5.2}$$

Both terms are divergent. To cancel the singularities, various methods are known, all based on the same idea of subtracting an auxiliary cross section from the real corrections. This is done in a way that $d\sigma^A$ has the same singular behavior as $d\sigma^R$. $d\sigma^A$ should be analytically integrable over the one-parton subspaces, causing the soft and collinear divergences. Finally, it can be combined with the virtual contribution to a finite correction. The NLO term can then be written as [58]:

$$\sigma^{NLO} = \int_{k+1} [(d\sigma^R)_{\varepsilon=0} - (d\sigma^A)_{\varepsilon=0}] + \int_k [d\sigma^V + \int_1 d\sigma^A]_{\varepsilon=0} \ . \tag{5.3}$$

---
[1]Monte Carlo generators apply stochastic methods and calculate physical processes and effects based on random numbers and statistical probabilities.
[2]NLOJET++ additionally contains matrix elements for proton-antiproton or e^+e^- collisions (among other processes).

The Catani-Seymour dipole subtraction method is one possible implementation for the numerical calculation of the NLO cross section.

For each randomly chosen element of the phase-space dx_1, dx_2 the according cross section $d\sigma$ is calculated in NLOJET++, representing the weight of the integration over the whole phase-space [2]. The output of the calculation is saved as a binary file. The results are then normalized and the statistical errors are calculated [59].

To cluster the final state partons according to the jet definition (in this case the exclusive k_T algorithm), the program FastJet[3] is used.

When observing 2-parton-events (where no emission of an additional parton takes place) the cross section of the LO (born) term is proportional to α_s^2 (see figure 5.1, left). 3-parton events are in leading order proportional to α_s^3 (see figure 5.1, right).

The NLO term includes some corrections of the cross sections. For 2-parton-events where either an emission of an additional parton or some virtual loop corrections appear, the cross section is proportional to α_s^3. Likewise the cross section of 3-parton-events is in NLO $\sigma \sim \alpha_s^4$ [2].

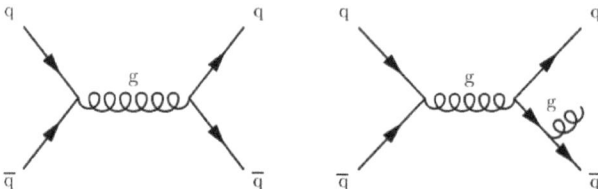

Figure 5.1: Left: LO process without the emission of an additional parton (the cross section is proportional to α_s^2). Right: LO process with an emission of an additional parton (the cross section is proportional to α_s^3) [2]

The parton distribution function (see chapter 2.5) CTEQ66M has been used in this analysis.

As an example, the figure 5.2 shows the p_T distribution of parton jets for 3-parton-events. Here, the k_T algorithm in the exclusive mode was forced to find exactly 3 jets with $p_T > 20$ GeV and $|\eta| \leq 2.6$. 10^8 events have been calculated, where NLOJET++ applies various iterations (like additional ISR or FSR) for each event. The black curve (full) represents the sum of the leading order term (LO or born) and the next-to-leading order term (NLO). In some bins full is not exactly LO+NLO, because for all three curves different events have been calculated (a simultaneous calculation of the same event in LO and NLO is not possible). The NLO term gives a correction of about 10%. In spite of the high statistics of 10^8 events, the phase-space is not sampled often enough, because the NLO curve drops down steeply (compared to the other curves). Furthermore, there are even some negative entries at $p_T > 200$ GeV.

5.2 PYTHIA and Underlying Event Models

The Monte Carlo event generator **PYTHIA** (version 6.4.24) [21] has been used to study the influence of hadronization effects and the Underlying Event. First of all, the program **PYTHIA** is shortly described, including the simulated subprocesses. Then, some UE models of **PYTHIA** are presented.

---

[3]FastJet is a fast implementation of several k_T algorithms for pp collisions. It is partly based on tools and methods from the computational geometry community as well as an original implementation of the e^+e^- algorithms. The jet implementations can be accessed via a plugin mechanism of the FastJet interface [60].

5.2. PYTHIA and Underlying Event Models

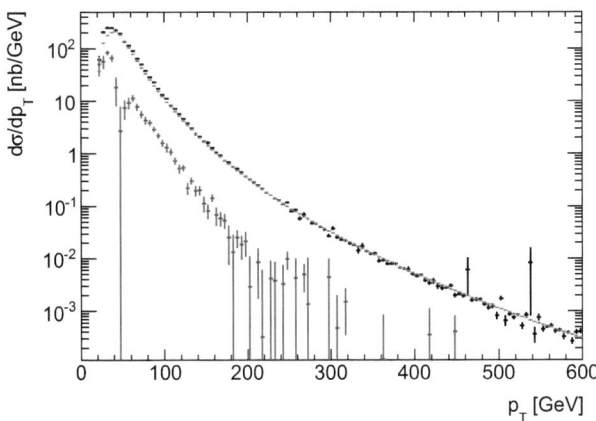

Figure 5.2: p_T distribution in LO (born), NLO and full (LO+NLO)

5.2.1 PYTHIA

The Monte Carlo generator PYTHIA is commonly used to simulate hadronic processes. With the help of random generators quantum mechanical fluctuations of the particle formations and decays are simulated depending on calculated matrix elements. As a leading order generator, PYTHIA only uses the matrix elements in first order of the perturbation theory of QCD for the simulation. The hadronization of the generated partons is described by the string model (see chapter 2.3).
PYTHIA has implemented a set of different PDFs. In addition, it is possible to link external PDF libraries to the program. In order to be consistent with NLOJET++, the PDF CTEQ66M has been applied using the PDF sets from the Les Houches Accord PDF Interface LHAPDF (version 5.8.4) [61]. The jet clustering is done via a C++ implementation of the k_T clustering algorithm, described in [45], as it can easily be included into PYTHIA. The implementation of the exclusive k_T algorithm shows small differences compared to FastJet, which handles the case $d_{min} = d_{kB}$ in an optimized way, but hardly changes the results.
In this analysis, QCD events coming from pp collisions at a center-of-mass energy of $\sqrt{s} = 7$ TeV have been studied (see table 5.1 for the chosen subprocesses[4]). For further information on the subprocesses, see [21].

The parameter CKIN(3) defines the minimal transfer of transverse momentum of the colliding particles via a cut-off in the phase-space. In table 5.1, the value of p_{Tmin} was set to 20 GeV. This means that relatively low energetic $2 \rightarrow 2$ collisions are generated[5].

The influence of the detector is not simulated in PYTHIA. This means that all created particles are detected and appear in an event. Particles very close to the beam or in regions with large η are e.g. not detected in real experiments due to the detector geometry. Nonetheless, PYTHIA offers plenty of opportunities to study high energetic processes, like the influence of the UE.

---

[4] In bracket the number of the according subprocess (MSUB) is shown. It can be either on (**1**) or off (0).
[5] Additionally to these parameters, PYEDIT has been set to **1** in order to get only stable final state particles.

Parameter	Subprocess
MSUB (11,1)	$f + f' \to f + f' (QCD)$
MSUB (12,1)	$f + \bar{f} \to f' + \bar{f}'$
MSUB (13,1)	$f + \bar{f} \to g + g$
MSUB (28,1)	$f + g \to f + g$
MSUB (53,1)	$g + g \to f + \bar{f}$
MSUB (68,1)	$g + g \to g + g$
CKIN (3,20)	$p_{T\text{min}}$ at hard $2 \to 2$ scattering

Table 5.1: Subprocesses for the generation of the hard scattering process

5.2.2 Underlying Event Models

In order to simulate the Underlying Event, models are needed as perturbation theory cannot be applied at this low energy range. Consequently, PYTHIA uses UE tunes, i.e. sets of parameters. These UE models can be added to the simulation of the hard scattering process. The final state particles (those from the UE as well as from the hard scattering) are then combined to jets.
Several of those UE tunes are available in PYTHIA. The according parameters have been tuned to experimental data from predecessor experiments and extrapolated to the high center-of-mass energies of LHC.
As none of the tunes describes the experimental data perfectly and nobody knows which tune is the best approximation of the UE, three different, most recent tunes have been investigated in this analysis: **ATLAS MC09c** [62], **AMBT1** [63] (already including data gathered at LHC) and **PERUGIA10** [64].

These models use the new p_T-ordered time-like final state parton shower (MSTP(81)=21). A description of the parameters can be found in [21]. A short explanation of some parameters and the default values are shown in table 5.2.
The parameters of the UE tunes are chosen to describe the Underlying Event from CDF[6] Run1 and Run2 and DØ[7] at Tevatron.
To study the Underlying Event at CDF, regions in the η-Φ-space have been analyzed, which are sensitive to the UE (see figure 5.3). The direction of the leading calorimeter jet ($jet\#1$) serves as a reference of the azimuthal angle. $\Delta\Phi = \Phi - \Phi_{jet\#1}$ stands for the relative angle between a charged particle and the direction of $jet\#1$. Perpendicular to the plane of the hard scattering the *transverse region* ($60° < |\Delta\Phi| < 120°$) is defined, which is sensitive to the UE.
To optimize the tunes, only charged particles with $p_T > 0.5$ GeV inside $|\eta| < 1$ have been used. The jet reconstruction has been done with a cone algorithm (see chapter 4.2) of $R = 0.7$ ($|\eta(jet\#1)| < 2$).
Two classes of events can be distinguished:

- **Leading jet events** are events with no further restrictions for $jet\#2$ and $jet\#3$.

- **Back-to-back events** are a special case of the first group, where two jets with $p_T > 15$ GeV appear, being almost back-to-back ($|\Delta\Phi| > 150°$) with $p_T(jet\#2)/p_T(jet\#1) > 0.8$. Here, $jet\#1$ lies in the "toward" region, whereas $jet\#2$ is inside the "away" region.

The transverse regions are separated (according to the number of charged particles) into transMAX

---
[6]Collider Detector at Fermilab
[7]DØ is a detector at Fermilab. The name comes from its location on the accelerator ring.

5.2. PYTHIA and Underlying Event Models

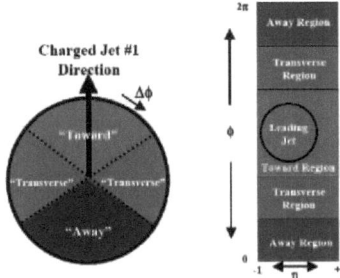

Figure 5.3: Definition of the transverse region [26]

(i.e. region with the highest

$$E_{Tsum} = \sum_i E_{T,i} \qquad (5.4)$$

of the particles) and transMIN (i.e. region with the smallest particle E_{Tsum}), to separate the hard (ISR and FSR) from the soft (beam remnant) component (see figure 5.4). The hard part can then be found in the transMAX region and the soft part in the transMIN region [26, 65, 66].

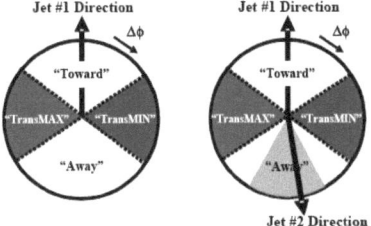

Figure 5.4: Definition of transMax and transMin [26]

The model, where the UE is described by regions in the η-Φ-space, which are distant to the influence of the leading jets, is called "swiss cheese model", as the UE is cut out via round cone jets.
When using data from LHC, the "swiss cheese model" has also been used to describe the UE, with the difference that the anti-k_T algorithm has been applied.
To calibrate the UE models, the parameters from table 5.2 have been tuned to fit the UE data from these regions. The modified parameters for the three UE models can be found in table 5.3.
The particular tunes are shortly described in the following:

- The parameters of **ATLAS MC09c** (adjusted by the ATLAS collaboration) employ the p_T-ordered parton shower with the MRST LO parton distribution functions. They are tuned to describe charged particle multiplicity distributions in minimum bias events of proton-antiproton collisions at 630 GeV and 1.8 TeV [62].

- **AMBT1** was tuned by the ATLAS collaboration and uses the PDF MRST LO. The parameters have been fit to ATLAS UE data and charged particle densities at 0.9 and 7 TeV. Additionally, CDF Run1 UE analyses and Z p_T-distributions, Run2 minimum bias and DØ Run2 dijet angular corrections have been used. The main tuning parameters are multiple parton interactions and color reconnections [63].

- **PERUGIA10** has been adjusted by Peter Skands [67]. To tune the soft QCD part, minimum bias data from proton-antiproton collisions from Tevatron and CERN have been used. It is an alternative to PERUGIA0 with more FSR off ISR, more beam remnant breakup and a higher production of $s\bar{s}$-quark-pairs [64].

In this analysis, PYTHIA has been used to study the UE, because the UE models describe the UE in data better than HERWIG [55].

5.3 HERWIG

In order to study the systematic uncertainties due to hadronization effects (see chapter 8.2.5), the Monte Carlo event generator HERWIG (version 6.510) [55] has been used to simulate events at parton and hadron level. The program is written in Fortran (a new version of HERWIG is also available in C++, labelled HERWIG++).

HERWIG offers a broad range of physical processes, including lepton-lepton, lepton-hadron or hadron-hadron scattering. The focus of the program is a detailed simulation of QCD parton showers. These showers are branching processes. The branchings are ordered in an angle from a maximum to a minimum value, which is determined by a cutoff [55].

HERWIG uses the cluster model to account for the hadronization (see chapter 2.3). In order to simulate the Underlying Event, the program JIMMY [68] can be linked to HERWIG.

In this analysis, hard QCD processes have been simulated on parton as well as on hadron level, using CTEQ66M from the external PDF set LHAPDF (version 5.8.4) [61].

The output is delivered in the HEP standard common block, labelled HEPEVT [55]. This data is converted in order to be readable with PYTHIA, where the jet reconstruction is done, using a C++ implementation of the k_T clustering algorithm [45].

Parameter	Default	Description
MSTP 51	7 (CTEQ 5L)	PDF set
MSTP 52	1	PDF set internal (=1) or pdflib (=2)
MSTP 3	2	QCD switch for choice of Λ_{QCD}
PARP 62	1 GeV	ISR IR cutoff
MSTP 64	2	ISR α_s type
PARP 64	1	ISR renormalization scale prefactor
MSTP 67	2	ISR coherence option for 1st emission
PARP 67	4	ISR Q_{max}^2 factor
MSTP 68	3	ISR phase-space choice
MSTP 70	1	ISR regularization scheme
MSTP 72	1	ISR scheme for FSR off ISR
PARP 71	4	FSR Q_{max}^2 factor for non-s-channel processes
PARJ 81	0.29 GeV	FSR Λ_{QCD}
PARJ 82	1 GeV	FSR invariant mass cut-off
MSTP 33	0	inclusion of K-factors
MSTP 81	1	UE model
PARP 82	2 GeV	UE IR cutoff at reference energy scale
PARP 89	1800 GeV	UE IR cutoff reference energy scale
PARP 90	0.16	power of energy-rescaling
MSTP 82	4	UE hadron transverse mass distribution
PARP 83	0.5	UE mass distribution parameter
PARP 84	0.4	UE mass distribution parameter
MSTP 88	1	beam remnant composite scheme
MSTP 89	1	beam remnant color connection scheme
PARP 79	2	beam remnant composite x enhancement
PARP 80	0.1	beam remnant breakup suppression
MSTP 91	1	beam remnant primordial k_T distribution
PARP 91	2 GeV	beam remnant primordial k_T width $< \lvert k_T \rvert >$
PARP 93	5 GeV	beam remnant upper cut-off for primordial k_T
MSTP 95	1	FS interaction color (re-)connection model
PARP 78	0.025	FS interaction color reconnection strength
PARP 77	0.0000	FS interaction color reco high-p_T damping strength

Table 5.2: Default parameters in PYTHIA. k_T is the transversal momenta evolution scale [21].

Parameter	ATLAS MC09c	AMBT1	Perugia10
MSTP 51	20650 (MRST2007LO)	20650 (MRST2007LO)	7 (CTEQ 5L)
MSTP 52	2	2	1
MSTP 3	2	2	1
PARP 62	1.0000	1.0250	1.0000
MSTP 64	2	2	3
PARP 64	1.0000	1.0000	1.0000
MSTP 67	2	2	2
PARP 67	4.0000	4.0000	1.0000
MSTP 68	3	3	3
MSTP 70	0	0	2
MSTP 72	1	1	2
PARP 71	4.0000	4.0000	2.0000
PARJ 81	0.2900	0.2900	0.2600
PARJ 82	1.0000	1.0000	1.0000
MSTP 33	0	0	0
MSTP 81	21	21	21
PARP 82	2.3150	2.2920	2.0500
PARP 89	1800	1800	1800
PARP 90	0.2487	0.2500	0.2600
MSTP 82	4	4	5
PARP 83	0.8000	0.3560	1.5000
PARP 84	0.7000	0.6510	0.4
MSTP 88	1	1	0
MSTP 89	1	1	0
PARP 79	2.0000	2.0000	2.0000
PARP 80	0.1000	0.1000	0.1000
MSTP 91	1	1	1
PARP 91	2.0000	2.0000	2.0000
PARP 93	5.0000	10.0000	10.0000
MSTP 95	6	6	8
PARP 78	0.2240	0.5380	0.0350
PARP 77	0.0000	1.0160	1.0000

Table 5.3: Parameters of different UE tunes. In addition to these parameters, Perugia10 sets some specific Λ_{QCD} values via MSTU(112)=4, PARU(112)=0.1920, PARP(1)=0.1920, PARP(61)=0.1920 and PARP(72)=0.2600. Furthermore, some fragmentation parameters are set for all three UE tunes [21].

5.4 ATHENA

ATHENA[8] [56] is an implementation of a framework for high energy physics with the name GAUDI [70]. It has been specified for the ATLAS experiment (originally, GAUDI has been developed for LHCb). This object-oriented ATLAS software framework is designed to process and reconstruct real data and to perform physics analyses.

Additionally, simulated data from Monte Carlo generators can also be processed. For this reason, the geometry and the behavior of the ATLAS detector components are simulated as good as possible. To simulate the detector responses and the impact of the detector material on the final state particles, the program GEANT4 [71] is used.

Analyses with ATHENA can run on local machines as well as on the data & computing grid. For the user analysis, a Python file (commonly known as *joboption*) is needed to control and configure ATHENA. This file is read by the application manager and allows the interactive modification of diverse parameters. The joboption e.g. constitutes which triggers or jet algorithms are used when running on data.

In this analysis, ATHENA has been used to analyze both real data and fully simulated data from Monte Carlo generators. This simulation is done in several steps (see figure 5.5). First of all, scattering events are simulated by MC generators like PYTHIA [21], HERWIG [55], Alpgen [72], or MC@NLO [73]. The generated particles are stored in the HepMC format and are then modified by GEANT4. This package simulates the detector material as well as the magnetic field and furthermore includes effects like multiple scattering, the loss of energy and photon conversion. Afterwards, the expected detector responses (like pulses or drift times) are calculated based on the GEANT4 hits during the digitization. As the detector effects are included, the simulated data is at this state comparable to real data. Instead of running the full simulation, digitization and reconstruction chain, the program ATLFAST [74] can be used, approximating these steps by smearing the 4-vectors from Monte Carlo generators according to the detector resolutions (taken from fully simulated events). The computation time is in this way reduced by several orders of magnitude [5]. (In this analysis, only fully simulated data have been used.)

In the next step particle tracks and calorimeter clusters are reconstructed, yielding the four-momenta. These are then stored as candidates for physics objects (e.g. electrons or jets) in ESDs[9] and AODs[10] (being derived from ESDs).

For typical user analyses, centrally produced ntuples are used. These are then analyzed with programs like Root [75]. The ntuples are derived from AODs and contain only observables, which are important for the specific physics channel that is investigated (each working group has its own specific ntuples).

Unfortunately, at this state, it is not possible anymore to reconstruct jets with a certain (non-standard) algorithm. Thus, this analysis runs directly on AOD files (and not on the ntuples of the Standard Model working group), which can only be accessed within the ATHENA framework. As the k_T algorithm in the exclusive mode is not stored in the AODs, the clustering process has to be re-run using local calibrated topo-clusters as an input (see chapter 4.4). The jet reconstruction is done by the program FastJet[11] using the jet reconstruction package [76] of ATHENA.

The output of the AOD analysis is then analyzed with Root.

ATHENA is constantly under development resulting in several available releases. In this thesis, ATHENA release version 15.6.10.6 has been used [77].

---

[8]The ATHENA framework is written in C++ (with parts released in Fortran) and Python [69].

[9]Event Summary Data: reconstructed information, containing enough objects (like the original calorimeter clusters) to redo the reconstruction. ESDs are used for calibration and optimization of jet reconstruction algorithms.

[10]Analysis Object Datas contain less information than ESDs. The focus of the stored data lies on physics objects (like four-momenta or reconstruction quality). The reconstruction of jets with non-standard algorithms is still possible.

[11]The program FastJet has also been used for the jet reconstruction in NLOJET++.

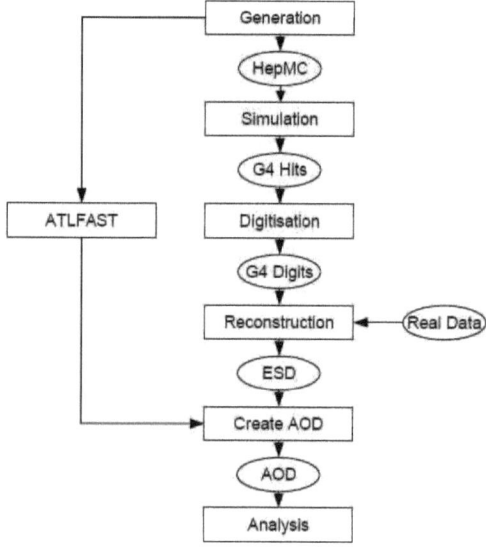

Figure 5.5: Monte Carlo simulation with the ATHENA framework (based on [5])

Chapter 6

Differential 2-Jet-Rate

In this chapter the differential 2-jet-rate is investigated. This observable has been used in this analysis to determine the strong coupling constant α_s. Starting with a short motivation, the differential 2-jet-rate at LO and NLO is studied, using the program NLOJET++. In the next section, real data from the ATLAS detector at the LHC are analyzed. For this reason the datasets used, run list and triggers are presented and jet cleaning cuts are provided. Then, 3-jet-events and 4-jet-events are compared in order to separate them. The real data are then compared to fully simulated data from PYTHIA as well as to calculations from NLOJET++. Finally, simulations from PYTHIA at parton level are checked against calculations from NLOJET++.

6.1 Motivation

There are several different ways to determine the strong coupling constant. α_s is included into every observable where jets are involved. These observables can be jet cross sections, ratios of jet cross sections or the internal structure of jets. Inclusive single jet- and multi jet-events (see e.g. [78]) can for example be used for α_s studies. Event shapes (often measured via the variable thrust) are also correlated to α_s. In addition, α_s can be determined from hadronic τ decays [79], from Z decays or from lattice QCD.

The transition parameter from 3→2 jets from the Durham [80] jet algorithm has been used for the α_s determination in former collider experiments, especially at e^+e^- colliders. This flip-parameter is equivalent to the measurement of the ratio of trijet to dijet events. In this way, the theoretical uncertainties can be reduced as many of them cancel in the ratio.

The k_T algorithm in the exclusive mode is based on the Durham jet algorithm and therefore also allows to access the flip-values from 3→2 reconstructed jets (see chapter 4.3.2). These flip-values are not very sensitive to the jet-energy-scale and hence allow the measurement of α_s at an early stage of the experiment. Figure 6.1 shows the values of α_s, measured with different methods.

By combining many different measurements, the world average was set to $\alpha_s(M_Z) = 0.1184 \pm 0.0007$ [81].

6.2 Studies with NLOJET++

The program NLOJET++ (see chapter 5.1) calculates parton production at leading and next-to-leading order. The idea of the α_s determination in this thesis is to perform a fit for α_s of calculations from NLOJET++ to real data. In order to get the different α_s dependencies, NLOJET++ has been used to calculate 2-parton-events at NLO, 3-parton-events at LO as well as NLO and 4-parton-events at LO.

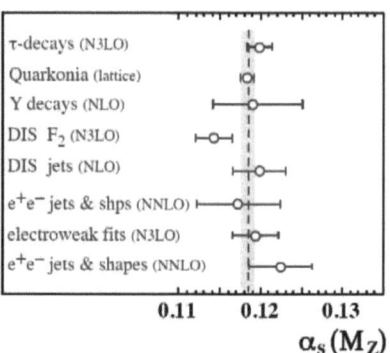

Figure 6.1: Summary of measurements of $\alpha_s(M_Z)$. The vertical line and yellow band mark the final 2009 world average value of $\alpha_s(M_Z) = 0.1184 \pm 0.0007$ determined from these measurements [81].

6.2.1 d_{23} Distributions of 2-Parton-Events

First of all, 10^8 events with 2 partons ($p_T > 20$ GeV and $|\eta| < 2.6$) at a center-of-mass energy of 7 TeV have been calculated at NLO. A leading order distribution of d_{23} is obviously not available when simulating 2-parton-events, because the flip-value from 3 to 2 is zero, as there are no events at LO with more than 2 jets. In this case the full calculation is therefore identical with the calculation of the NLO term. The figures 6.2 show the distribution of d_{23} of 2-parton-events at NLO.

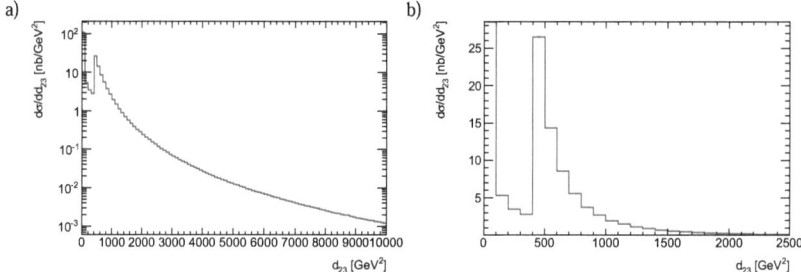

Figure 6.2: d_{23} distribution of 2-parton-events a) logarithmical and b) non-logarithmical in NLO. Due to the large statistics, the error bars are very small.

From 400 GeV^2 on the distribution decreases smoothly to higher values of d_{23}. The jet clustering algorithm (see chapter 4.3) and the cut on the jets' transverse momenta cause a cut at $d_{23} = 400$ GeV^2. As a minimum transverse momentum of at least 20 GeV is demanded for all jets, the value $d_{kB} = p_{Tk}^2$ of the exclusive k_T algorithm is at least 400 GeV^2. The values $d_{23} < 400$ GeV^2 are hence all coming from $d_{kl} = min(p_{Tk}^2, p_{Tl}^2) \times R_{kl}^2$, with an R_{kl} value smaller than 1. As FastJet handles this region slightly differently than the k_T jet implementation used for the simulations from PYTHIA, a cut has been set at $d_{23} = 400$ GeV^2 and events with $d_{23} < 400$ GeV^2 have been ignored for the α_s determination. The peak at zero is due to virtual corrections at NLO where no additional parton in the final state appears. The flip-value from 3 to 2 for these events with only 2 jets in the final state is zero.

6.2. Studies with NLOJET++

In this distribution, many different values of α_s are included as there is no distinction between different values of Q. To account for the Q dependency of α_s, Q has been approximated by the p_T of the leading jet. The values of d_{23} are allocated to intervals of the leading jet p_T of the associated event[1]. Six different p_T intervals with a width of 20 GeV have been chosen, covering a p_T region from 20 GeV up to 140 GeV (see figures 6.3).

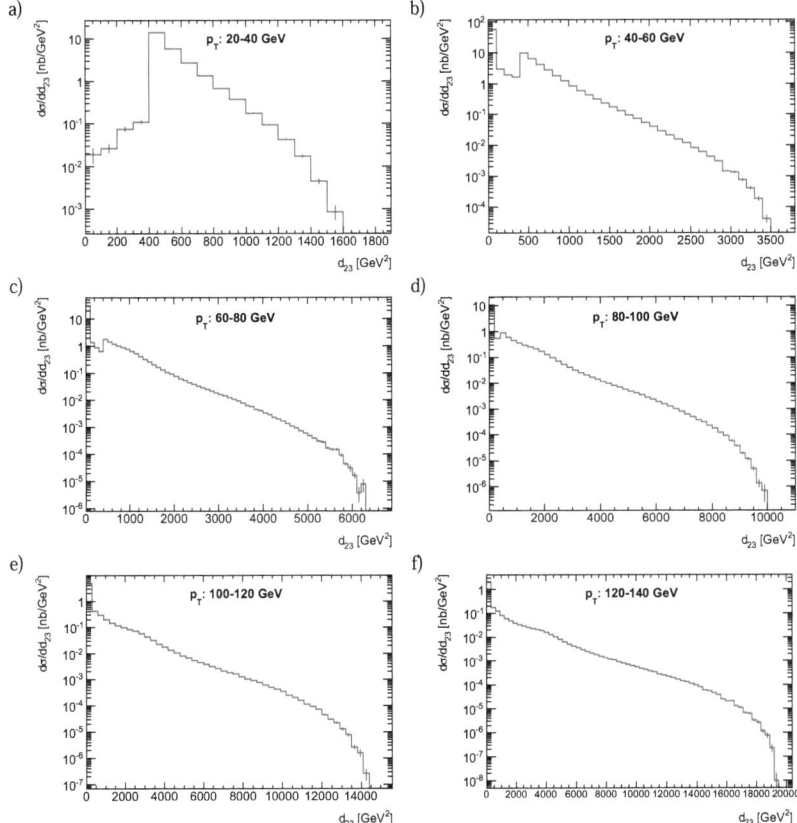

Figure 6.3: d_{23} distributions of 2-parton-events, separated in p_T intervals of the leading jet in NLO calculation

The bin width amounts to 100 GeV^2, 200 GeV^2 and 300 GeV^2 for figures a) to c), for figure d) and for figures e) and f), respectively.
In consistence with the clustering algorithm ($d_{kB} = p_{Tk}^2$), the maximum values of d_{23} relate to p_T^2 of the right boundary of the according interval (e.g. 1600 GeV^2 for figure a) with a p_T interval from 20 to 40 GeV). In this extreme case, all jets have more or less the same transverse momentum (e.g. in figure a) the third jet, being crucial for the flip-value, must have a p_T of around 40 GeV).
Higher transverse momenta therefore lead to higher values of d_{23}.

---
[1] As the transverse momentum of the leading jet has been used to separate the d_{23} distributions, p_T^2 of the leading jet has been chosen for the hard scale.

6.2.2 d_{23} Distributions of 3-Parton-Events

The same calculations have been done for 3-parton-events. Here, also the calculation at LO provides a d_{23} distribution, as there are 3 jets in the final state and therefore flip-values from 3 to 2 jets available. This distribution is almost identical to the NLO distribution of 2-parton-events (see figures 6.4).

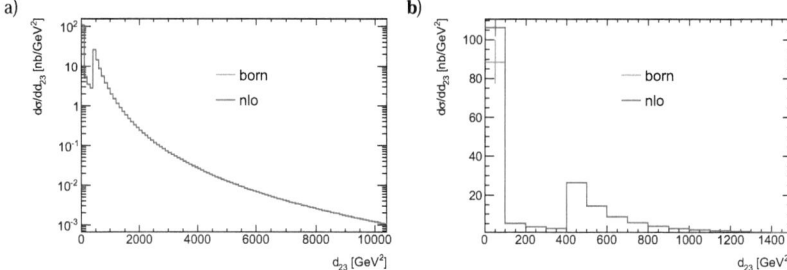

Figure 6.4: d_{23} distribution of born (3-parton-events) and nlo (2-parton-events) a) logarithmical and b) non-logarithmical. Due to the large statistics, the error bars are very small.

The real correction of the NLO term with 2 partons is commensurate with the integral over the born matrix element of 3 jets in the final state [58]. The good match of the above curves indicates very small virtual corrections in NLOJET++[2]. Some virtual terms of NLO from 2-parton-events, compensating divergencies of the cross section, are noticeable for very small values of d_{23} [2].

Figures 6.5 compare the full, the LO and the NLO calculation of the d_{23} distributions of 3-parton-events. The large statistical fluctuations of the NLO distribution required the calculation of 3×10^8 NLO events.

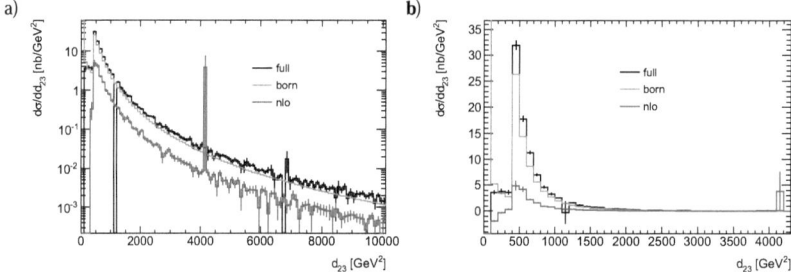

Figure 6.5: d_{23} distribution of 3-parton-events a) logarithmical and b) non-logarithmical in full, LO and NLO calculation

The NLO term increases the cross section by about 10%. The peak of the NLO distribution at 4100 GeV^2 is due to a lack of statistics. As the error of this value is quite large, this bin has been aligned to the curve when fitting the strong coupling constant (see chapter 8.1).
The calculation of the full theory, i.e. LO "plus" NLO, using 10^8 events shows a negative entry at 1100 GeV^2. Negative entries should not appear, as LO "plus" NLO should in total result in

---
[2] The curves are still in good agreement when divided into p_T intervals of the leading jet.

6.2. Studies with NLOJET++

positive entries. A reason could be that the phase-space is still not sampled often enough, or that the transition between different regions in phase-space is not smooth enough in NLOJET++. The author of NLOJET++ has been notified about this, but could not provide a (better) explanation. As the full distribution is not used for the α_s fit, this is not crucial.

The d_{23} distributions divided into p_T intervals of the leading jet are shown in figures 6.6.

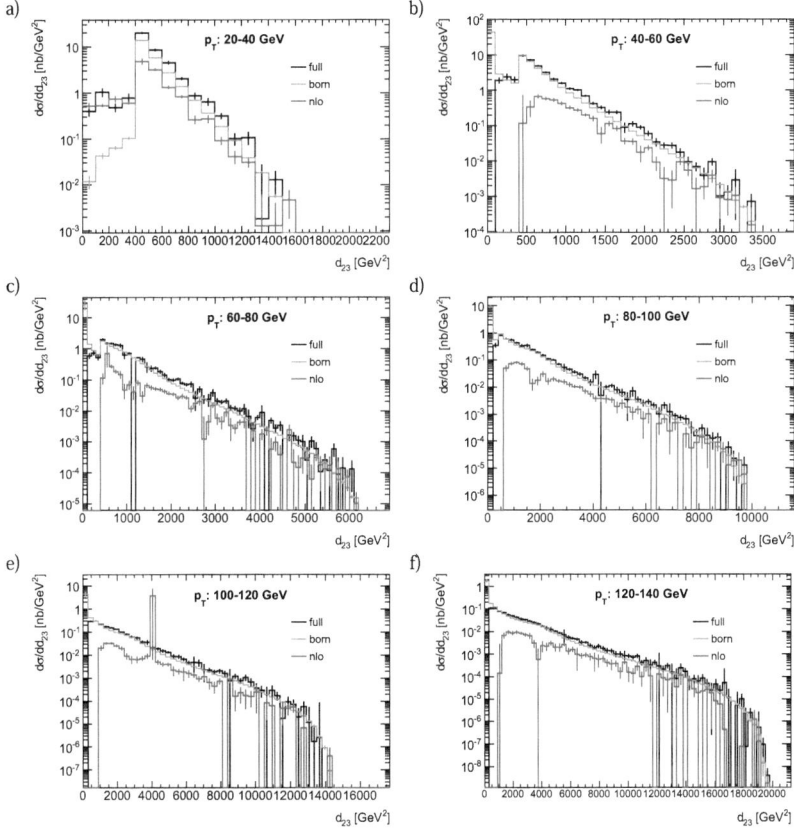

Figure 6.6: d_{23} distributions of 3-parton-events in different p_T intervals in full, LO and NLO calculation. The p_T of the leading jet is inside the according interval.

The curves of the LO and NLO distributions are different (the d_{23} distributions at NLO are e.g. flatter than the distributions at LO). This means that higher orders of α_s change the shapes of the d_{23} distributions. The flip-values are therefore sensitive for the strong coupling constant. Thus, the shape of the distributions can be used to determine α_s.

6.2.3 d_{23} Distributions of 4-Parton-Events

NLOJET++ calculates 4-parton-events just at LO. Therefore, in figures 6.7 only the LO distribution of d_{23} is shown.

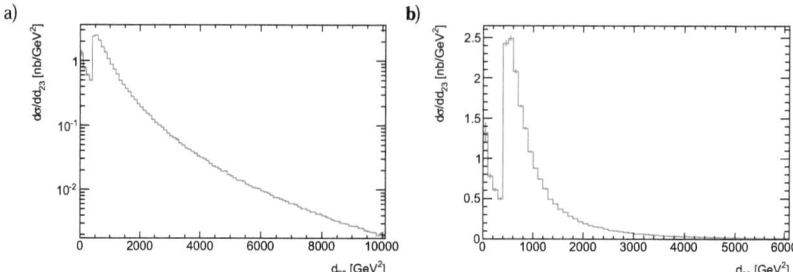

Figure 6.7: d_{23} distribution for 4-parton-events a) logarithmical and b) non-logarithmical in LO calculation. Due to the large statistics, the error bars are very small.

The curve is very smooth and shows no outliers. The cross section of this distribution amounts to about 25% of the distribution for 3-parton-events at LO. The d_{23} distribution is again divided into intervals of p_T of the leading jet (see figures 6.8).

The d_{23} distributions of the leading order calculations of 3- and 4-parton-events have been compared to the distributions from Alpgen [72]. The results from NLOJET++ and Alpgen have been in good agreement to each other.

6.2. Studies with NLOJET++

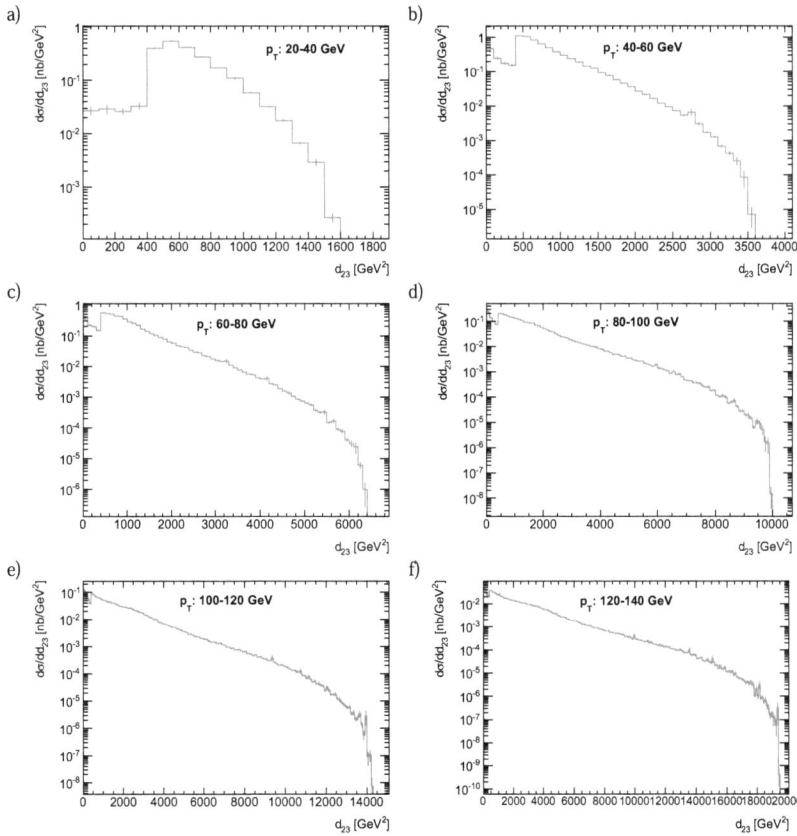

Figure 6.8: d_{23} distributions of 4-parton-events in different p_T intervals in LO calculation. The p_T of the leading jet is inside the according interval.

6.2.4 Comparison between BORN and FULL "minus" NLO

The d_{23} distributions of 3-parton-events have been further investigated. When simulating the born term everything (e.g. internal α_s, PDFs) is calculated at leading order, whereas these internal values are calculated at next-to-leading order for the NLO calculation. To determine the influence of these internal parameters it has been tested if the born term is consistent with the full calculation, where the NLO term is subtracted. This comparison has neither been done for 2-parton-events (as the NLO term is in this case identical with full), nor for 4-parton-events (where the born term represents the full calculation due to a missing NLO implementation in NLOJET++).

The figures 6.9 show the born distribution compared to full "minus" the NLO term.

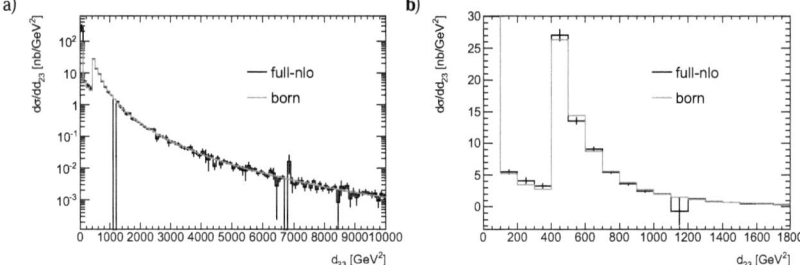

Figure 6.9: Comparison between born and full "minus" nlo for 3-parton-events a) logarithmical and b) non-logarithmical.

The distributions are also divided in p_T intervals of the leading jet (see figures 6.10).

The curves are in good accordance and match within the statistical fluctuations. Thus, the internal calculations at LO and NLO yield only small differences within statistical uncertainties.

6.2. Studies with NLOJET++

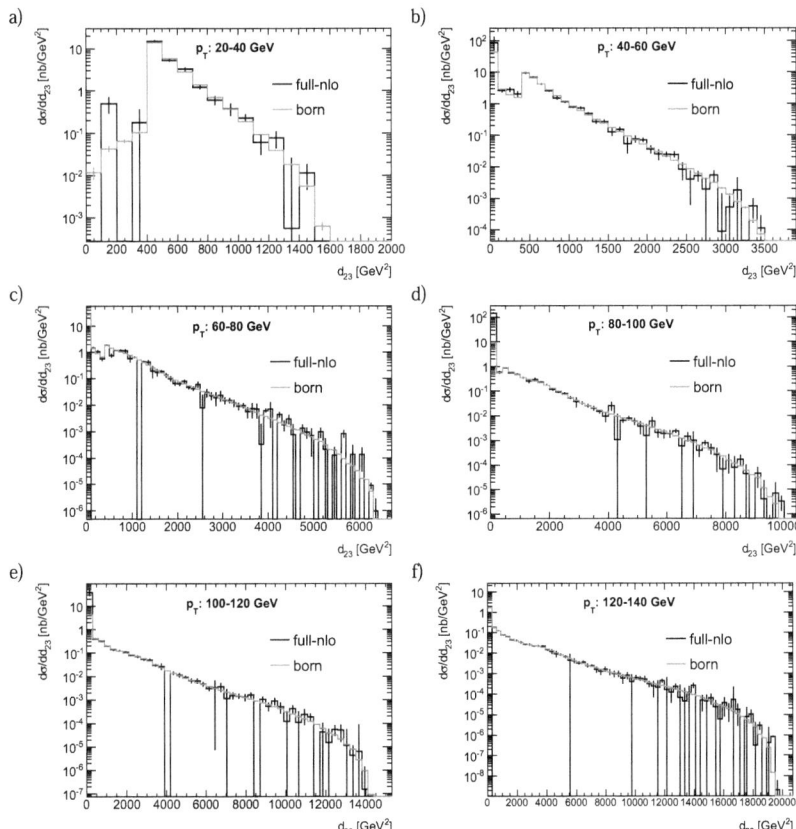

Figure 6.10: Comparison between born and full "minus" nlo in different p_T intervals. The p_T of the leading jet is inside the according interval.

6.3 Real Data Analysis

This section presents the datasets used, the good run list and the triggers, which have been applied to data. Then, some jet cleaning variables are studied and compared to the anti-k_T jet algorithm. Finally, the d_{23} distributions are compared to the d_{34} distributions, in order to separate 3-jet-events from 4-jet-events.

6.3.1 Datasets

ATHENA has been used to analyze the datasets from run periods A-F, shown in table 6.1.

Run Period	Run Number	# Events	Dataset Name
A	152166-153200	2020106	data10_7TeV.periodA.physics_L1Calo.PhysCont.AOD.repro04_v01
B	153565-155160	17438115	data10_7TeV.periodB.physics_L1Calo.PhysCont.AOD.repro04_v01
C1-C2	155228-156682	14043136	data10_7TeV.periodC.physics_L1Calo.PhysCont.AOD.t0pro04_v01
D1-D6	158041-159224	94578192	data10_7TeV.periodD.physics_L1Calo.PhysCont.AOD.t0pro04_v01
E1-E7	160387-161948	45598178	data10_7TeV.periodE.physics_JetTauEtmiss.PhysCont.AOD.t0pro04_v01
F1-F2	162347-162882	34937674	data10_7TeV.periodF.physics_JetTauEtmiss.PhysCont.AOD.t0pro04_v01

Table 6.1: Datasets from run periods A-F used for this analysis. The numbers are taken from AMI [82].

These data have been gathered by ATLAS, starting data taking at a center-of-mass energy of 7 TeV on March 30th 2010. The runs from periods G to I are not used in this analysis, as different triggers are applied for these datasets. Figure 6.11 shows the total integrated (cumulative) luminosity (periods A-I) versus day, being delivered by LHC (green) and recorded by ATLAS (yellow) at $\sqrt{s} = 7$ TeV for all pp runs in 2010 [83].

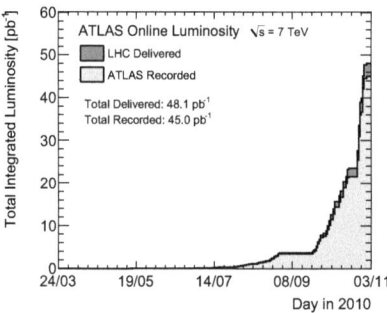

Figure 6.11: The total integrated luminosity versus day [83]

6.3. Real Data Analysis

A good run list has been applied to exclude events and luminosity blocks considered as bad (e.g. due to not properly working detector components). For this analysis, the good run list merged_grl_PeriodABCDEFGHI_152166.206-167776.546_SMjet_nomuon_7TeV_noHLT_A-F.xml has been used. This is a combined list of several run periods at 7 TeV [84], provided by the Standard Model group.

The good events have been processed on the data & computing grid (see chapter 3.3) and the results further analyzed. Unfortunately, several events got lost due to problems with the data & computing grid (e.g. ATHENA crashes because of refused connections), which could not be solved.

The jet triggers L1_J15 and L1_J30 have then been applied to the remaining events. Table 6.2 shows the official integrated luminosity for each run period for the two jet triggers.

Period (nb^{-1})	L1_J15	L1_J30
A	0.38	0.38
B	8.1	8.1
C1	7.2	7.2
C2	1.3	1.3
D1	27.5	27.5
D2	32.9	32.9
D3	32.9	32.9
D4	79.5	79.5
D5	28.0	28.0
D6	97.1	97.1
E1	32.5	139.0
E2	91.8	92.5
E3	90.2	237.2
E4	1.7	88.7
E5	2.0	129.7
E6	1.8	153.9
E7	12.5	161.4
F1	3.2	400.2
F2	5.6	293.2
Sum	556.0	2010.6

Table 6.2: Integrated luminosity for each run period [84]

In table 6.3 the numbers of analyzed events before and after the triggers are presented. The trigger L1_J15 is prescaled[3] in run periods E and F, whereas only period F is prescaled for the trigger L1_J30. Hence, in this analysis, mainly run periods A to E with the trigger L1_J30 (with an integrated luminosity[4] of around 700 nb^{-1}) have been analyzed.

---

[3]The prescale factors reduce the number of events examined by the triggers.
[4]Due to the above reasons, the analyzed integrated luminosity is much smaller than the 1317.2 nb^{-1} from run periods A to E in table 6.2.

6.3.2 Jet Cleaning

Before jets can be further analyzed they have to be cleaned from detector effects. Hence, the jets have to comply with certain conditions. As described in chapter 4.5.3, there are several cuts available, deciding if a jet is bad, ugly or good.

However, these cuts have been optimized for the anti-k_T jet algorithm (see chapter 4.3.3) and not for the k_T jet algorithm in the exclusive mode. Thus, some of these cut variables are studied for both algorithms. It has to be mentioned that this comparison is just qualitatively possible, because both algorithms have significant differences in their definitions. For this study, the datasets from period B have been used exemplarily to reduce processing time while having sufficient statistics. No special trigger has been demanded, meaning that all events firing any trigger have been studied.

Figure 6.12 shows the p_T distributions of the leading jets for the different jet algorithms (with a minimum jet p_T of 7 GeV). The differences of the distributions are expected. The anti-k_T algorithm with a radius of 0.4 (blue curve) finds more low energetic jets and less high energetic jets than the other algorithms. With a radius of 0.6 (black curve), more particles are assigned to a jet, leading to higher jet momenta. Finally, the k_T jet algorithm in the exclusive mode (red curve) finds more jets with high p_T. As the algorithm is forced to find 3 jets per event, it clusters more particles to a jet, resulting in large jets with a high p_T.

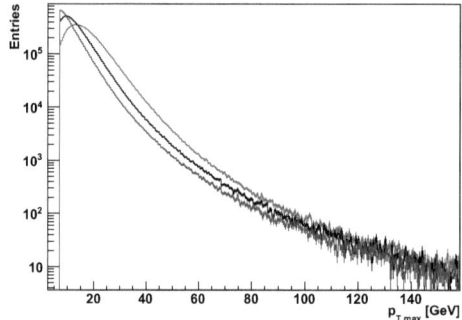

Figure 6.12: p_T distributions of the leading jets, being clustered by the exclusive k_T and the anti-k_T algorithm (radius 0.6 and radius 0.4)

The comparison of the jet cleaning variables is done exemplarily for some important variables, like f_{HEC}, f_{EM} and $|t|$:

First of all, f_{HEC} is investigated. Sporadic noise bursts in the hadronic end cap calorimeter (HEC) are the most common reason for mis-reconstructed jets. These noise bursts deposit most of their energy in single calorimeter cells, often with some entries in the neighboring cells. For this reason they can be excluded if a large fraction of the jet-energy is found in the HEC, accompanied by a low number of cells, accounting for at least 90% of the jet-energy [54]. These quantities are used to cut-off the jets from noisy HEC clusters.

The distributions of f_{HEC} (see figure 6.13) have approximately the same shape. They start at negative values and have a maximum at around 0. The entries in this area are due to the energy resolution of the HEC. Then, the curves decline to 1. With the cut-off value of $f_{HEC} > 0.8$ (for the loose bad jet definition) the anti-k_T algorithms classify more jets as potentially bad than the exclusive k_T jet algorithm. This does, however, not mean that these jets are indeed bad. Sporadic noise bursts would

6.3. Real Data Analysis

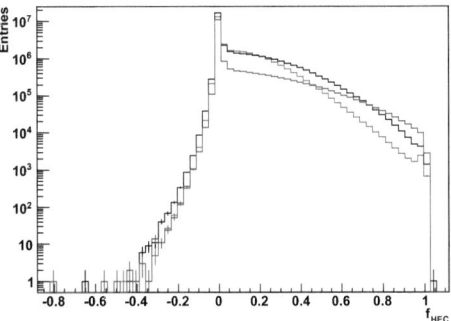

Figure 6.13: f_{HEC} distributions of jets clustered by the exclusive k_T and the anti-k_T algorithm (radius 0.6 and radius 0.4)

be seen in the distributions as peaks. No boosts can be observed at $f_{HEC} < 0.8$, meaning that all algorithms are safe for this noise.

In addition to the HEC, also the electromagnetic calorimeter can be affected by noise bursts, which are, however, not very frequent. Jets from these coherent noise bursts are characterized by a large reconstructed energy in the electromagnetic calorimeter and a bad quality of the calorimeter reconstruction. The quality is evaluated via the difference in the sampling of the measured pulse and a reference pulse. The latter is used for the reconstruction of the cell energy [54]. Bad jets are tagged if the fraction of the jet-energy from bad-quality calorimeter cells is above a certain value and the fraction of reconstructed energy in the electromagnetic calorimeter in general $f_{EM} > 0.95$ (for the loose definition) or $f_{EM} > 0.90$ (for the tight definition) (see figure 6.14).

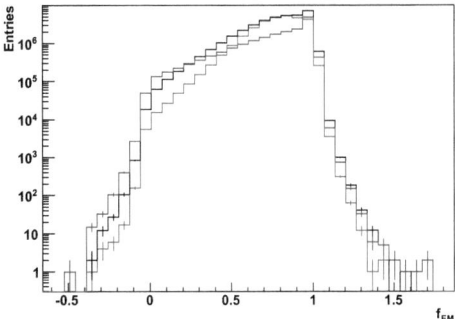

Figure 6.14: f_{EM} distributions of jets clustered by the exclusive k_T and the anti-k_T algorithm (radius 0.6 and radius 0.4)

The distributions of f_{EM} have basically a similar shape: rising from negative values due to the energy resolution of the EM the curves reach a maximum at around 1, dropping down to about 1.5. The results from the anti-k_T algorithm with radius 0.4 differs most from the other algorithms. Still, the anti-k_T algorithms with different cone sizes use the same jet cleaning cuts. Thus, the cuts are also fine for the exclusive k_T jets, where the agreement with anti-k_T (radius 0.6) is better than the agreement of the two anti-k_T jets. As there are no peaks at $f_{EM} < 0.9$, the algorithms are safe for noise bursts in the EM with this cut-off value.

Another criterion for the jet classification is the out-of-time energy deposition in the calorimeter (see figure 6.15), where the jet time is defined with respect to the event time. These energy depositions can result from photons produced by cosmic ray muons. The energy-squared-weighted cell time should be within two beam bunch crossings, otherwise the jets are classified as bad [54].

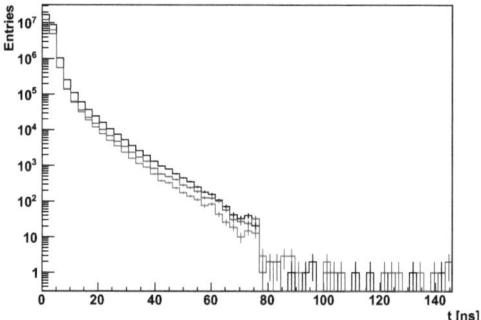

Figure 6.15: Jet time distributions of jets clustered by the exclusive k_T and the anti-k_T algorithm (radius 0.6 and radius 0.4)

The shape of the jet time distributions for the exclusive k_T and the anti-k_T algorithms look quite similar. A cut on this variable should therefore have similar results in all cases.

The above jet cleaning variables show a similar behavior for the exclusive k_T and the anti-k_T jet algorithms respectively. This is not surprising since the cleaning variables are mainly based on bad calorimeter cells, which are more or less independent of the used jet algorithm. Due to the same behavior, the jet cleaning cuts from the anti-k_T algorithm have been used for this analysis and events with one or more bad jets removed.

In addition to these jet cleaning cuts, all three jets have been required to have $p_T > 20$ GeV and $|\eta| < 2.6$. Only events fulfilling those criteria have been accepted and analyzed. Table 6.3 shows the number of events for each run period, before and after the triggers, jet cleaning and further cuts.

6.3. Real Data Analysis

Period	A	B	C	D	E	F
# events (no trigger)	726599	1.116e+07	1.173e+07	2.728e+07	2.733e+07	2.737e+06
after jet cleaning	610174	8.586e+06	8.988e+06	2.451e+07	2.621e+07	2.663e+06
$p_T > 20$ & $|\eta| < 2.6$	5803	132685	162256	1.573e+06	3.797e+06	501056
after L1_J15	6110	171915	213497	2.873e+06	4.726e+06	17220
after jet cleaning	5632	163025	203759	2.800e+06	4.609e+06	16865
$p_T > 20$ & $|\eta| < 2.6$	1186	31921	40527	589356	973440	3691
after L1_J30	993	25326	31354	397366	2.289e+06	255830
after jet cleaning	774	22427	28702	388258	2.261e+06	253056
$p_T > 20$ & $|\eta| < 2.6$	303	8828	11246	152952	901353	100932

Table 6.3: Number of events before and after the triggers and cuts

6.3.3 Separation from 4-Jet-Events

The differential jet-rates depend on d_{cut}. At smaller values of d_{cut}, the 4-jet-rate dominates, whereas the fraction of the 3-jet-rate becomes more and more important with increasing d_{cut} values.
In order to separate the 3-jet-rate from the 4-jet-rate, the distributions of d_{23} and d_{34} (i.e. the flip-value from 4 to 3 jets) are compared in figures 6.16 - exemplarily for run periods A-E (trigger L1_J30)[5].

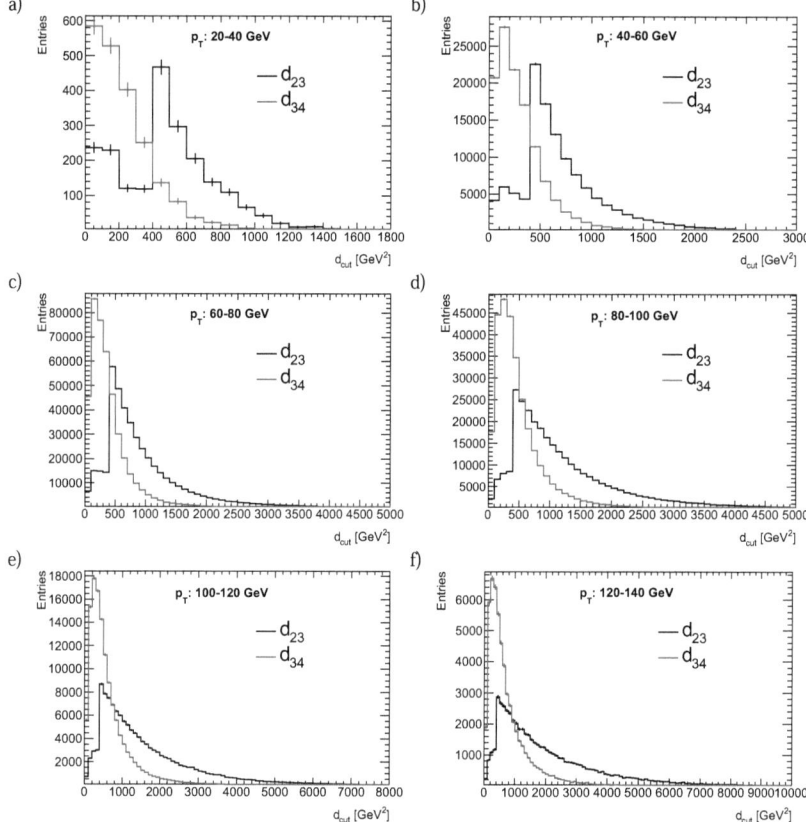

Figure 6.16: d_{23} and d_{34} distributions for run periods A-E (trigger L1_J30). Due to the large statistics, the error bars are very small from figure b) onwards.

The d_{34} distributions decline steeper than the d_{23} distributions. For small values of d_{cut} (which is either the flip-value from 4 to 3 jets or the flip-value from 3 to 2 jets), d_{34} is dominant, whereas d_{23} has more entries at higher d_{cut} values.
In order to separate the unwanted fraction of d_{34} from the d_{23} distributions (in this analysis, only 3-jet-events are investigated), the integrals (starting from different values of d_{cut}) of these distributions

---
[5]Period F (trigger L1_J30) is prescaled and therefore not included into the plots.

6.3. Real Data Analysis

are compared. Thus, the ratio between the integrals is calculated:

$$R(d_{cut}) = \frac{\int d_{23}}{\int d_{34}}, \tag{6.1}$$

where $\int d_{ij}$ stands for $\int_{d_{cut}}^{\infty} \frac{d\sigma}{dd_{ij}} dd_{ij}$.
The ratios are shown in figures 6.17.

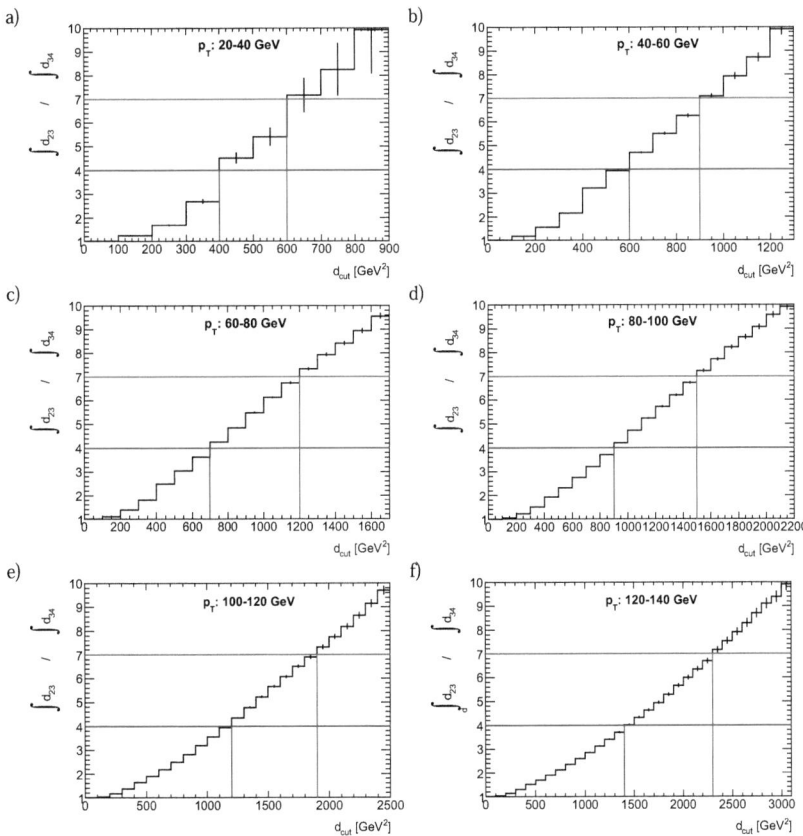

Figure 6.17: Ratio of $\int d_{23}$ to $\int d_{34}$

At higher values of d_{cut} for every d_{34} value a corresponding d_{23} value can be found. Therefore A_0 can be defined as $A_0 = \int d_{23} - \int d_{34}$. The errors are then calculated (with $\Delta A_0 = \sqrt{A_0}$ and

$\Delta \int d_{34} = \sqrt{\int d_{34}}$):

$$\Delta \frac{\int d_{23}}{\int d_{34}} = \sqrt{\left(\frac{\int d_{23}}{\int d_{34}} - 1\right)^2 \times \left[\left(\frac{\Delta A_0}{A_0}\right)^2 + \left(\frac{\Delta \int d_{34}}{\int d_{34}}\right)^2\right]} = \sqrt{\left(\frac{\int d_{23}}{\int d_{34}} - 1\right)^2 \times \left(\frac{1}{A_0} + \frac{1}{\int d_{34}}\right)} = \sqrt{\frac{\int d_{23}}{(\int d_{34})^2} \times \left(\frac{\int d_{23}}{\int d_{34}} - 1\right)}.$$
(6.2)

The blue line at $R = 4$ in figures 6.17 indicates an impurity of 20%, i.e. $\int d_{34}$ represents a fraction of $\frac{1}{5}$ to the total distribution. Consequently, $\int d_{23}$ has a fraction of $\frac{4}{5}$. The ratio $\int d_{23}:\int d_{34}$ of $\frac{4}{5}:\frac{1}{5}$ is fulfilled if the division of $\int d_{23}$ by $\int d_{34}$ (which is done in the above figures) yields a value of 4. Accordingly, the red line at $R = 7$ stands for an impurity of 12.5% (with a ratio $\int d_{23}:\int d_{34}$ of $\frac{7}{8}:\frac{1}{8}$). With rising values of R, the impurity (and therefore the systematic error) decreases - at the expense of statistics. The intersections of the lines with the R distribution yield the according d_{cut} value (see table 8.1 in chapter 8.1). The regions smaller than these d_{cut} values have been excluded when performing the α_s fit, as the fraction of d_{34} is too large in this area. The influence of the different impurities is studied in chapter 8.2.2.

6.4 Comparison to Simulations

In this chapter, real data (from run periods A-E) are compared to simulations. First of all, the d_{23} distributions from real data are compared to fully simulated PYTHIA dijet samples. Then, real data are checked against the calculation from NLOJET++. Finally, the simulation from PYTHIA at parton level is compared to the results of NLOJET++.

6.4.1 Comparison between Data and PYTHIA

The standard QCD samples, i.e. fully simulated PYTHIA dijet samples (see table 6.4), are studied in this chapter.
The samples have been processed on the data & computing grid (see chapter 3.3) with ATHENA, running the exclusive k_T jet algorithm (N=3) on truth particles as well as on LCTopo clusters[6]. As the application of the jet triggers L1_J15 and L1_J30 would decrease the number of events (especially in the smaller p_T intervals of the leading jet) to unusable small statistics, no special jet trigger has been used. Therefore, when comparing the samples to real data, no special jet trigger requirement has been applied to the real data in order to assure comparability.
Before the merging of the samples, they have been scaled according to their different cross sections and number of events.
Events with bad jets have been excluded and jets with $p_T > 20$ GeV and $|\eta| < 2.6$ required for both simulation and real data. A further cut has been set on the d_{23} value and events with $d_{23} < 400$ GeV^2 have been neglected. In this area, the fraction of d_{34} is too high. Moreover, the cut has been set to have the same conditions as for the calculations with NLOJET++ (see chapter 6.2.1).
The normalized d_{23} distributions are shown in figures 6.18.
In red, the d_{23} distribution of the truth jets is shown, whereas the green curve already takes detector effects into account[7]. Hence, the green and the black curve, representing real data, should be in good agreement. This is true for figures c) to f) - especially in the interesting p_T interval 80-100 GeV where the mass of the Z boson is located. However, when going to smaller energies, see e.g. figure a) and b), the curves are quite different. The data are not described very well by reconstructed simulation (which is not completely understood). When low energetic dijet events are forced by the jet algorithm to find

---
[6]Some failed jobs reduced the number of events of dataset 105009 to 1294186 and of dataset 105011 to 1288081.
[7]The green and the red curve are used in chapter 7.1 to estimate a correction for the jet-energy-scale.

6.4. Comparison to Simulations

Dataset	p_T [GeV]	σ [nb]	# Events	Dataset Name
105009	8-17	9752970	1399184	mc09_7TeV.105009.J0_pythia_jetjet.merge. AOD.e468_s766_s767_r1303_r1306
105010	17-35	673020	1395383	mc09_7TeV.105010.J0_pythia_jetjet.merge. AOD.e468_s766_s767_r1303_r1306
105011	35-70	41194.7	1398078	mc09_7TeV.105011.J0_pythia_jetjet.merge. AOD.e468_s766_s767_r1303_r1306
105012	70-140	2193.25	1397430	mc09_7TeV.105012.J0_pythia_jetjet.merge. AOD.e468_s766_s767_r1303_r1306
105013	140-280	87.8487	1397401	mc09_7TeV.105013.J0_pythia_jetjet.merge. AOD.e468_s766_s767_r1303_r1306
105014	280-560	2.32856	1391612	mc09_7TeV.105014.J0_pythia_jetjet.merge. AOD.e468_s766_s767_r1303_r1306
105015	560-1120	0.0338461	1347654	mc09_7TeV.105015.J0_pythia_jetjet.merge. AOD.e468_s766_s767_r1303_r1306

Table 6.4: PYTHIA dijet samples. The numbers are taken from AMI [82].

3 jets, the last step of the clustering routine is redeemed, resulting in a low transverse momentum of the 3rd jet (which leads to a small value of d_{23}). This jet is more likely to have only a smaller energy than the 3rd jets of events with originally more than 2 jets. All different types of jet events can be found in real data, resulting in higher energetic 3rd jets and therefore higher values of d_{23} compared to the PYTHIA dijet samples. This circumstance carries more weight at lower than at higher energies. At higher energies, also the 3rd jets from original dijet events have a high p_T. Thus, in events with high momentum transfers, the curves match well.

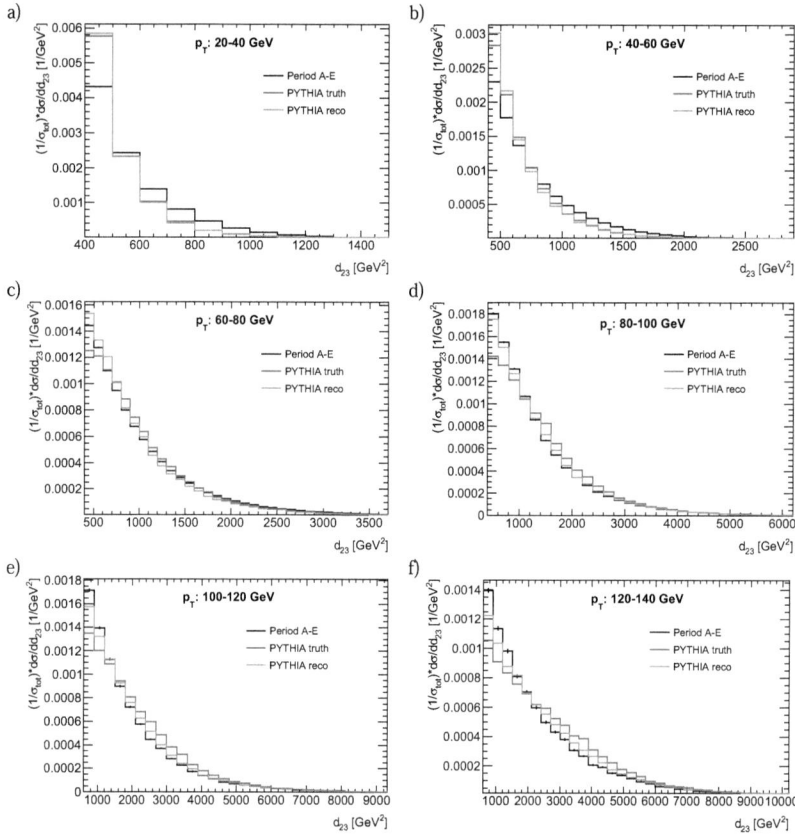

Figure 6.18: d_{23} distributions from data (periods A-E) and from simulation with PYTHIA with and without detector simulation. Due to the large statistics, the error bars are very small.

6.4.2 Comparison between Data and NLOJET++

To determine α_s, the full calculation of **NLOJET++** is split into **LO** and **NLO** terms, having different dependencies on the strong coupling constant. These terms are then compared to real data.
Therefore, in figures 6.19 the full calculation of **NLOJET++** (of 3-parton-events) is compared to real data (run periods A-E). The same cuts as in chapter 6.4.1 have been used for both distributions (as the calculation of **NLOJET++** has only been done on parton level without detector simulation, no jet cleaning cuts have been applied for **NLOJET++**). No special trigger has been used in both cases.

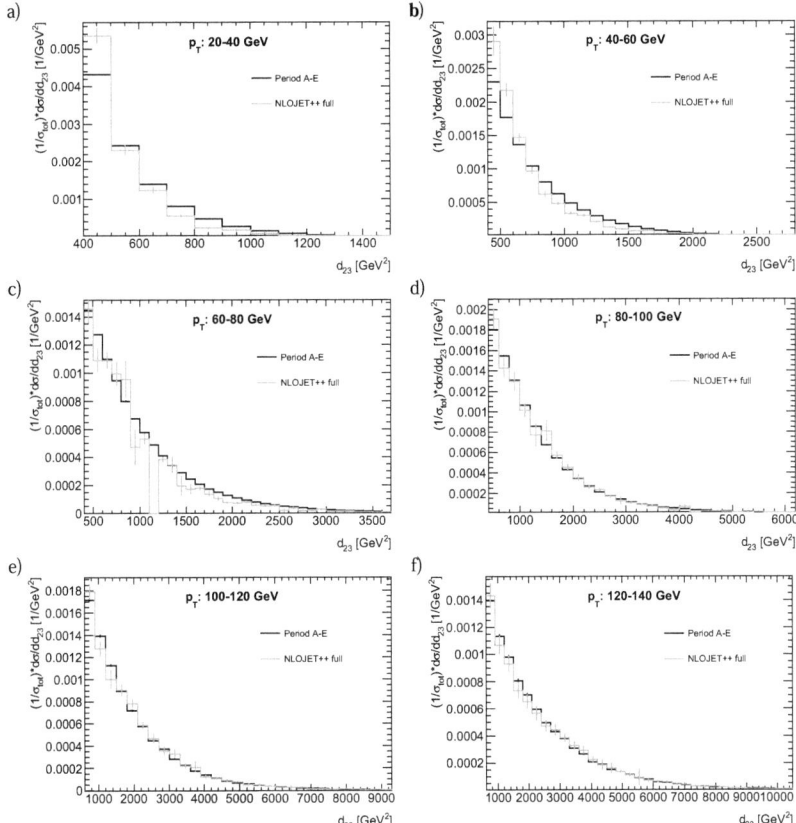

Figure 6.19: d_{23} distributions from data (periods A-E) and from simulation with NLOJET++ (full)

The curves match quite well from figure d) onwards, but are different in figures a) to c) for several reasons. The jet-energy-scale has e.g. not been corrected for the real data. In addition, no hadronization or Underlying Event effects had been taken into account for the **NLOJET++** simulation (these corrections are all done in chapter 7). These effects - above all the UE (see chapter 7.3) - lead to quite large deviations in regions where Q is small. With increasing Q, the impact of the UE declines. Hence, the curves are in good agreement at higher values of Q.

6.4.3 Comparison between PYTHIA and NLOJET++

In order to correct the calculation of NLOJET++ for hadronization effects and the influence of the Underlying Event, the calculations from NLOJET++ are corrected bin-by-bin. This is done in chapters 7.2 and 7.3. The correction factor for the hadronization is determined with PYTHIA (on generator level) by dividing hard scattering processes at hadron level by the hard scattering at parton level. The UE correction is also done with PYTHIA by dividing the hard scattering with UE by hard scattering without UE, both on hadron level.

Before the correction factors are evaluated, the parton levels of PYTHIA and NLOJET++ (born, 3-parton-events) are compared (see figures 6.20). 100,000 events have been simulated with PYTHIA[8]. Table 6.5 shows the CKIN(3) values (i.e. the minimal transfer of transverse momentum of the colliding particles via a cut-off in the phase-space), which have been set for the different p_T intervals of the leading jet. The values are chosen by subtracting twice the jet-energy-scale correction of about 5% from the left interval border.

p_T Interval	CKIN(3) value [GeV]
20-40 GeV	18
40-60 GeV	36
60-80 GeV	54
80-100 GeV	72
100-120 GeV	90
120-140 GeV	108

Table 6.5: CKIN(3) values used for the according p_T intervals of the leading jet

The distributions show some differences: the red curve is flatter than the black curve. Thus, PYTHIA finds more events with high values of d_{23} than NLOJET++ does. The differences can be explained by the different simulation parameters (PYTHIA uses e.g. the parameter CKIN(3), which can not be set in NLOJET++). Moreover, Pythia is leading order complemented by a leading logarithmic parton shower. A perfect agreement of PYTHIA and NLOJET++ is hence not expected. Although well matching curves would of course be favorable, PYTHIA can still be used to correct NLOJET++ for hadronization and UE effects. In this way, migration effects from one bin to the adjacent bin are compensated. Therefore, relative corrections in the according bins can be applied to NLOJET++.

---

[8] All stable and long lived particles have been considered, but excluding neutrinos.

6.4. Comparison to Simulations

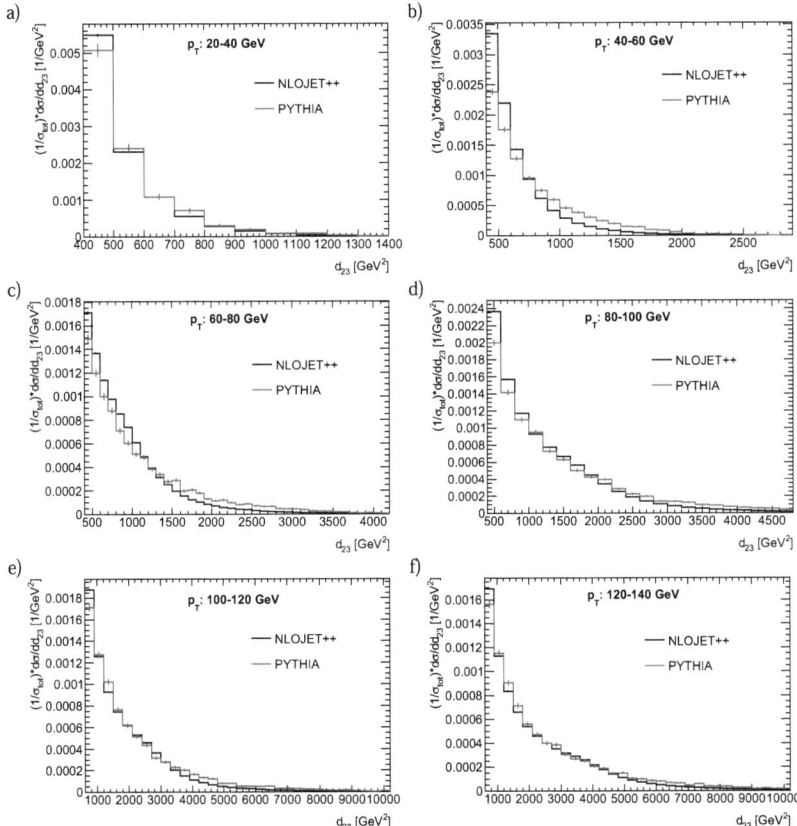

Figure 6.20: d_{23} distributions from NLOJET++ (born, 3-parton-events) and PYTHIA. The p_T of the leading jet is inside the according interval.

Chapter 7

Corrections

Before α_s is determined, some corrections are made to cope with the jet-energy-scale of real data and the missing hadronization and Underlying Event (UE) effects of NLOJET++. In order to compare real data to the calculations from NLOJET++, these adjustments are applied using bin-by-bin corrections. Moreover, a correction method is presented to directly correct data for the influence of the UE.

7.1 Jet-Energy-Scale

As already mentioned, the jet-energy-scale and the according uncertainty is measured in ATLAS only for the anti-k_T jet algorithm. The correction of detector effects has been done in this analysis for the exclusive k_T jet algorithm accordingly by comparing the d_{23} distributions of PYTHIA dijet samples, running the jet clustering on LCTopo clusters and on truth particles (see figures 6.18 in chapter 6.4.1).

In order to get correction factors, the d_{23} distributions of truth jets have been divided by the d_{23} distributions of reconstructed (i.e. LCTopo) jets:

$$H^i_{JES}(d_{23}) = \frac{d^i_{23,truth}}{d^i_{23,reco}} \ . \tag{7.1}$$

The correction factors in the according p_T intervals can be found in figures 7.1.
To take the JES into account, the data have been multiplied by the correction factors:

$$d_{23,data(JES\ corrected)} = H^i_{JES}(d_{23}) \times d_{23,data} \ . \tag{7.2}$$

At higher values of d_{23} the statistics get too low resulting in a fluctuation of the entries. Thus, no correction to data has been applied in figure c) for $d_{23} > 5500$ GeV^2, in figure d) for $d_{23} > 8000$ GeV^2, in figure e) for $d_{23} > 10000$ GeV^2 and in figure f) for $d_{23} > 12000$ GeV^2.

The result can be seen in figures 7.2 where the d_{23} distributions from periods A to E are shown before and after the correction of the JES.
The corrections are small in figures a) and b). However, in these two figures the mismatch of the d_{23} distributions between data and PYTHIA is quite large (see chapter 6.4.1). Moreover, the correction due to the shift of the energy-scale has potentially larger effects due to bin migrations between the p_T intervals of the leading jet, especially from the interval $p_T < 20$ GeV. Therefore, events with a leading jet's transverse momentum of less than 60 GeV are handled with care.

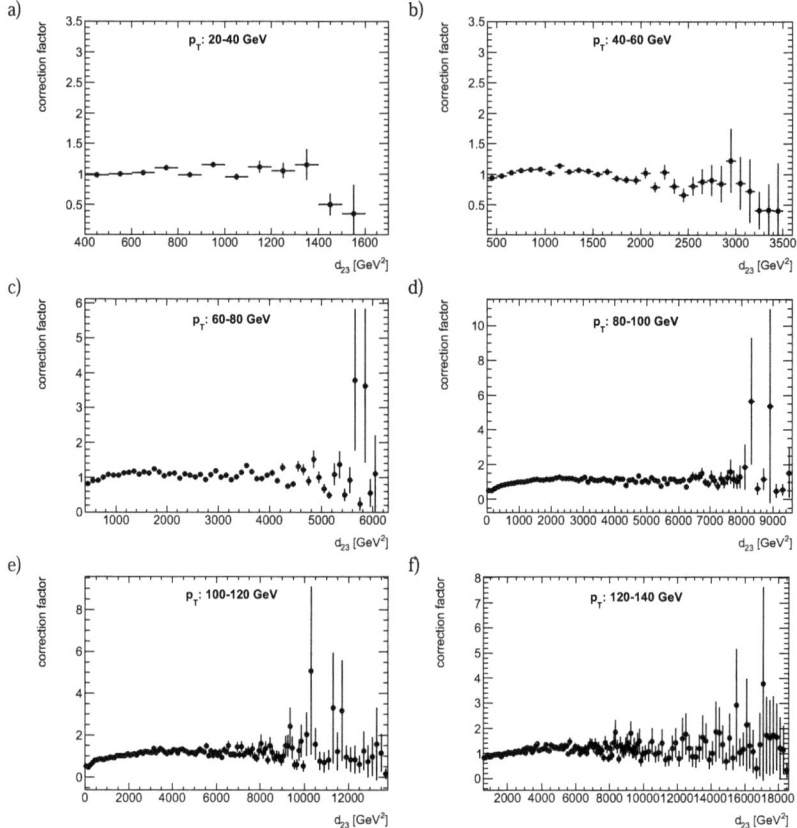

Figure 7.1: Correction factors for the JES

7.2 Hadronization

As already stated, NLOJET++ has unfortunately no hadronization model implemented and therefore calculates only at the parton level. In order to compare the results of NLOJET++ with real data, the influence of the hadronization has to be taken into account. The factorization of long- and short-distance physics are the reason why hadronization corrections to infrared safe observables can be estimated by running event generators at parton and hadron level [78]:

$$H^l_{hadr}(d_{23}) = \frac{d^l_{23,hadr}}{d^l_{23,part}} \quad . \tag{7.3}$$

In this study the program PYTHIA has been used to investigate hadronization effects on the differential 2-jet-rate, because with PYTHIA it is possible to simulate both the parton level and the hadron level. In order to calculate the systematic uncertainty due to the hadronization, also HERWIG has been used to simulate events on parton as well as on hadron level (see chapter 8.2.5).

7.2. Hadronization

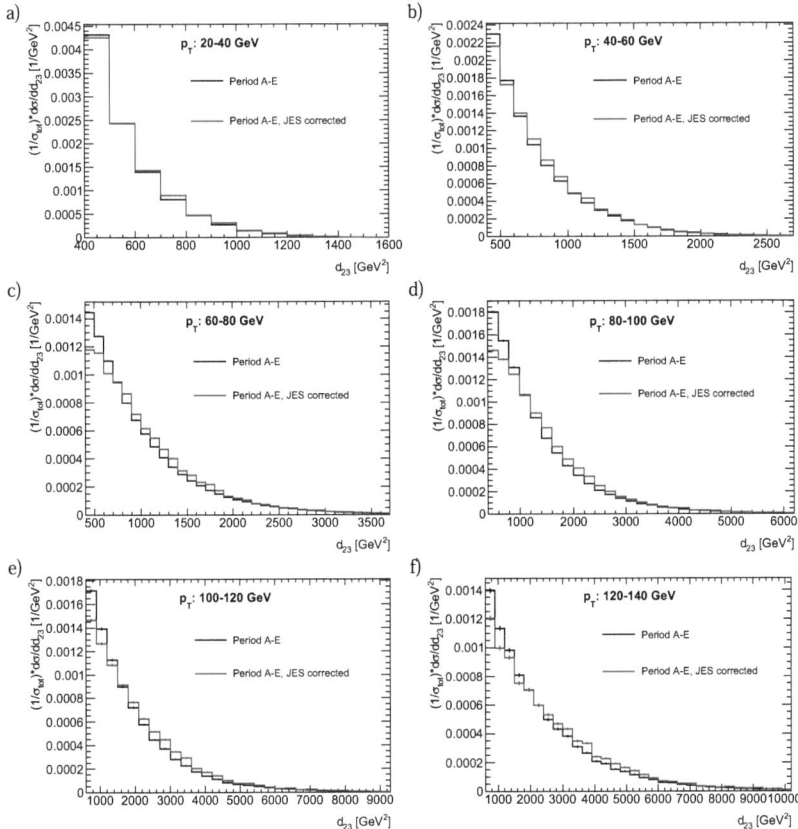

Figure 7.2: d_{23} distributions from periods A to E before and after the correction of the JES. The p_T of the leading jet is inside the according interval. Due to the large statistics, the error bars are very small.

The figures 7.3 show the d_{23} distributions before and after the hadronization in different p_T intervals of the leading jet.

At smaller regions of p_T larger discrepancies would be expected due to the $\frac{1}{Q}$ dependency of the hadronization [12]. However, the curves in figure a) match quite well because of the large bin width (in relation to the maximum d_{23} values). For high energetic jets hadrons are clustered around the direction of the partons. Hence, the difference between parton and hadron level and therefore the according corrections are not large for high energetic jets [2].

The difference between hadron and parton level is rather small in all examined p_T intervals of the leading jet. This means that the k_T algorithm in the exclusive mode is almost independent of hadronization effects, because of its infrared safeness.

Dividing the hadron by the parton distribution yields the correction factors $H_{hadr}^i(d_{23})$ (see figures 7.4).

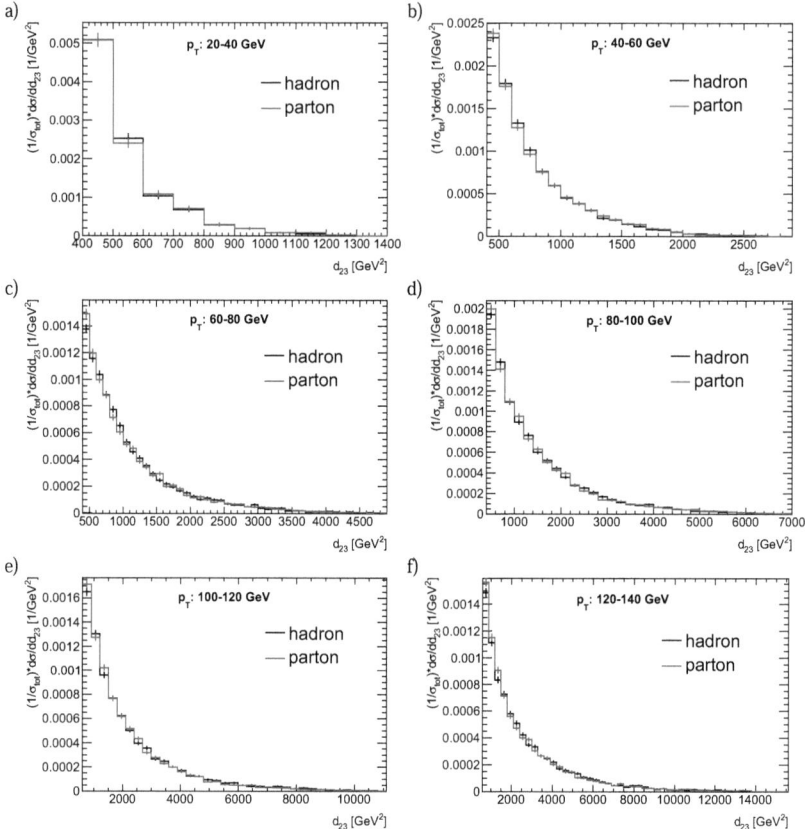

Figure 7.3: d_{23} distributions before and after hadronization. The p_T of the leading jet is inside the according interval.

When performing the α_s-fit (see chapter 8.1), the entries in the according bins are corrected by these factors:

$$d_{23,NLOJET++(hadr)} = H^i_{hadr}(d_{23}) \times d_{23,NLOJET++} \ . \tag{7.4}$$

Due to low statistics at high d_{23} values, the correction factors have been set to unity in figure b) for $d_{23} > 2000$ GeV^2, in figure c) for $d_{23} > 2800$ GeV^2, in figure d) for $d_{23} > 5500$ GeV^2, in figure e) for $d_{23} > 8100$ GeV^2 and in figure f) for $d_{23} > 10800$ GeV^2.

7.2. Hadronization

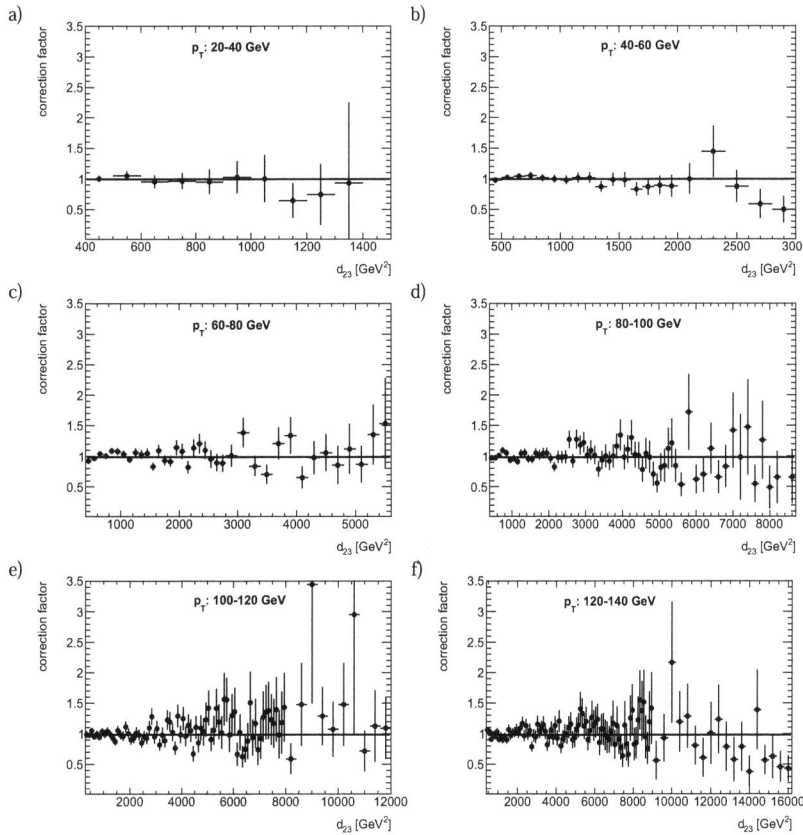

Figure 7.4: Hadronization correction. The black line represents a fit to the correction factors. The fit-values are shown in table 7.1.

p_T interval	Fit-value	Error
20-40 GeV	0.995	0.004
40-60 GeV	0.990	0.001
60-80 GeV	0.984	0.001
80-100 GeV	0.981	0.001
100-120 GeV	0.980	0.001
120-140 GeV	0.971	0.001

Table 7.1: Fit-values to the hadronization correction

7.3 Underlying Event

This chapter deals with the influence of the Underlying Event (UE) on the differential 2-jet-rate. As there is no UE model available describing the UE in a perfect way, three different UE models have been studied and compared to the hard scattering process without UE.
The program PYTHIA (on generator level) has been used to study the UE, applying the same CKIN(3) values (see chapter 5.2.1) as described in table 6.5.
The d_{23} distributions from hard scattering events without UE (labelled "hard") and with AMBT1 (labelled "hard+AMBT1"), PERUGIA10 (labelled "hard+PERUGIA10") and ATLAS MC09c (labelled "hard+ATLAS MC09c") are displayed in figures 7.5.

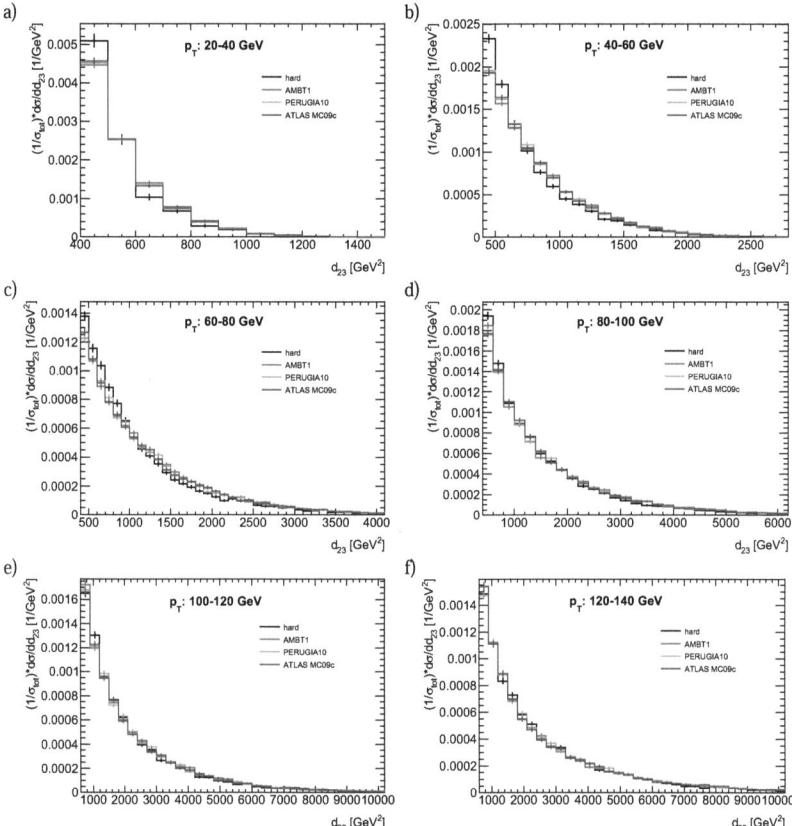

Figure 7.5: d_{23} distributions for "hard", "hard+AMBT1", "hard+PERUGIA10" and "hard+ATLAS MC09c". The p_T of the leading jet is inside the according interval.

The influence of the UE is quite large at small values of CKIN(3) and decreases with higher momentum transfers. By dividing the hard scattering process with UE by "hard" at hadron level, correction

7.3. Underlying Event

factors $H^l_{UE}(d_{23})$ are obtained:

$$H^l_{UE}(d_{23}) = \frac{d^l_{23,UE}}{d^l_{23,hadr}} \quad . \tag{7.5}$$

These correction factors are shown exemplarily for tune AMBT1 in figures 7.6.

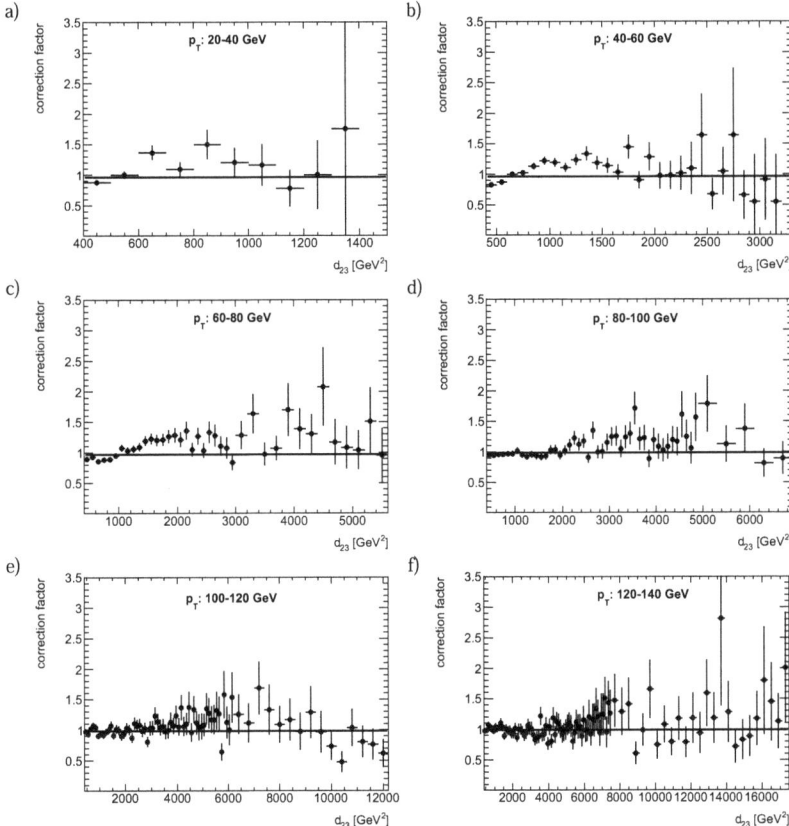

Figure 7.6: Underlying Event correction using tune AMBT1. The black line represents a fit to the correction factors. The fit-values are shown in table 7.2.

According to

$$d_{23,NLOJET++(UE)} = H^l_{UE}(d_{23}) \times d_{23,NLOJET++(hadr)} \quad , \tag{7.6}$$

the calculations from NLOJET++ are corrected bin-by-bin to cope with the influence of the UE.
Due to statistical fluctuations, no corrections apply, if the values of d_{23} are larger than a certain size. These values are shown in table 7.3 for the different UE tunes.
This bin-by-bin correction has been used in this thesis to account for the UE.
In addition, a correction method has been developed to directly correct data for the influence of the UE. This method is presented in the following chapter, exemplarily for the UE tune AMBT1.

p_T interval	Fit-value	Error
20-40 GeV	0.96	0.03
40-60 GeV	0.96	0.01
60-80 GeV	0.97	0.01
80-100 GeV	0.98	0.01
100-120 GeV	0.98	0.01
120-140 GeV	0.98	0.01

Table 7.2: Fit-values to the UE correction

p_T interval	AMBT1	PERUGIA10	ATLAS MC09
20-40 GeV	2000	2000	2000
40-60 GeV	2300	2200	2200
60-80 GeV	2900	3000	3300
80-100 GeV	5400	6200	5400
100-120 GeV	7900	7900	7900
120-140 GeV	8900	8900	8900

Table 7.3: Values of d_{23} in GeV^2 up to which d_{23} is corrected

7.4 Low-p_T Method

This section describes a correction method of k_T jet energies for the Underlying Event. This "low-p_T method" is based on [4], where the transverse momenta of jets and their constituents have been corrected for the contribution of the UE using the k_T algorithm in the exclusive mode. It is still under development and has only been tested so far on generator level using PYTHIA. Therefore, the immediate correction of real data for the impact of the UE is not possible yet. That is why the above bin-by-bin correction is used to consider the UE for the α_s determination.
Nonetheless, as the "low-p_T method" is very promising, it has been tested in this analysis on generator level for recent UE tunes. Additionally, the method has been applied to correct d_{23} values.

The k_T jet algorithm in the exclusive mode absorbs every particle it finds like a vacuum cleaner - no matter if the particle is coming from the UE or the hard scattering process. In this way, it is not possible to separate the unwanted UE particles e.g. geometrically, as it is done with cone or anti-k_T algorithms. For these jet algorithms clustering particles inside a fixed area in azimuthal angle φ and pseudorapidity η UE contributions are usually corrected by simply subtracting the average energy expected from particles of the UE measured in regions of the detector away from the hard scattering. The energy distribution is in that case obtained from regions away from the hard jet, like at 90° in φ and/or opposite η (see chapter 5.2.2). The k_T algorithm in the exclusive mode neither reconstructs jets of fixed area nor regular shape. For this jet algorithm a similar subtraction method has been preconceived for heavy ion collisions [85]. It is not possible to easily apply this method to proton proton collisions as the environment of particles from soft scatters is less dense.

The "low-p_T method" is inspired by measurements showing that the perturbative description can be extended to much lower scales than usually expected (see e.g. [12]). Moreover, experimental observations at HERA showed that the pertubative DGLAP evolution equations [86] effectually describe the PDF at very small values of Q (below 1 GeV^2) and x_{Bj} [87, 88].

Regarding a smooth transition between the perturbative description and non-perturbative effects, perturbation theory calculations approximatively describe parton parton collisions at essentially non-perturbative low scales. Thus, jets may be used to approximate non-perturbative contributions of the UE.

The "low-p_T method" therefore describes the UE by "low-p_T jets". These are the jets with the lowest transverse momentum in an event[1]. In this analysis, as the algorithm is forced to find exactly 3 jets in the final state, the low-p_T jet (comprising many particles from the UE) is identical with the 3rd jet in an event.

To verify this, the constituents of the 3rd jets from "hard+AMBT1" have been compared to particles from the UE. The simulation of solely UE (without the hard scattering process) is not possible in PTYHIA as the UE is intrinsically related to a hard scattering process (e.g. via color flow). Hence, the UE particles have been approximated by statistically subtracting the jet constituents from "hard" of the constituents from "hard+AMBT1" (in the following called "AMBT1"). The remaining particles represent indeed the UE simulated in PYTHIA. This has been tested by tracing the particles back to their origin.

As can be seen in figures 7.7, particles from "low-p_T jets" and from the UE ("AMBT1") have a very similar transverse momentum spectrum. This means that the "low-p_T jets" have the same composition as the UE and are therefore useful to approximate the contribution of UE particles.

---

[1] Particles from the UE may lead to the reconstruction of an additional (third) jet not related to a highly energetic gluon radiation.

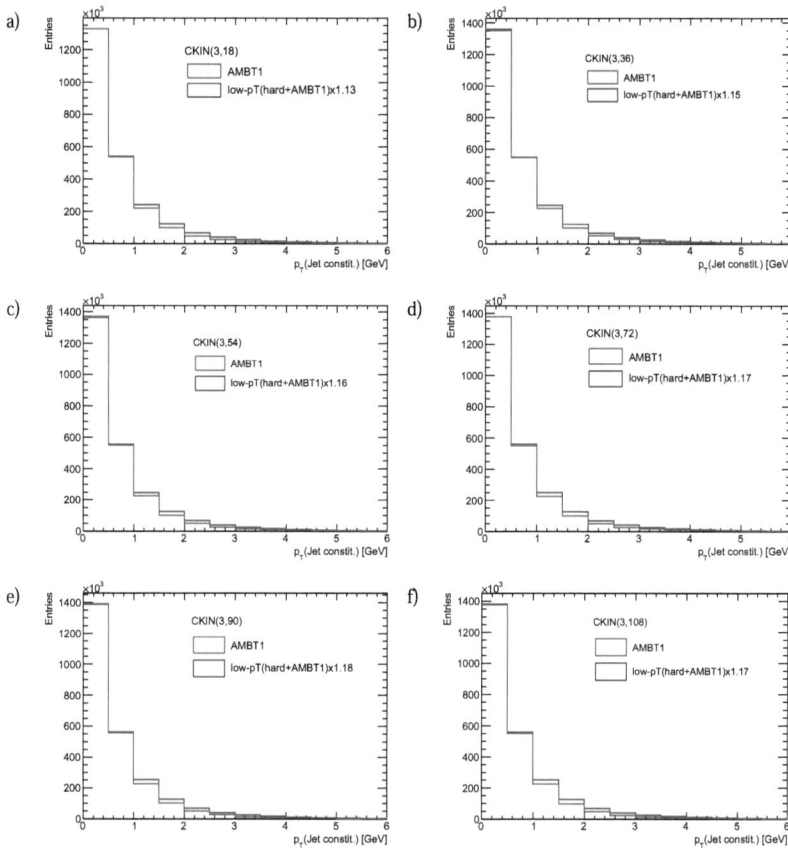

Figure 7.7: Particles' p_T of the UE (simulated with different values of CKIN(3)) and of the low-p_T jets (from CKIN(3,18) in all figures). Due to the large statistics, the error bars are very small.

7.4. Low-p_T Method

The comparison between "low-p_T jets" and UE has been done for many different UE tunes (see [4]). The good match of the different tunes indicates that the method is model independent.
Particles from the UE can be found in every jet (and not only in the 3rd jet), i.e. also in jets with high p_T. As the perturbative cross section of the UE compared to the "low-p_T jets" is in this area of the phase-space significantly higher, the constituents of the "low-p_T jets" have been scaled up in order to have the same particle content in regions of small transverse momenta. The scaling factors sf have values from 1.13 to 1.18.

It has to be mentioned that the "low-p_T jets" have been taken from the simulation of CKIN(3,18)[2] in all p_T intervals of the leading jet. If going to higher values of CKIN(3), the momentum transfer gets too high and consequently the jet constituents (also in the 3rd jet) too energetic. The "low-p_T jets" are then too energetic to describe the UE properly (at high energies, the 3rd jet is very likely to originate from highly energetic gluon radiation). As an example, "AMBT1" with CKIN(3,108) is described by the "low-p_T jet" (also from CKIN(3,108)) in figure 7.8.

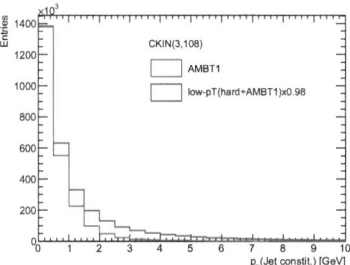

Figure 7.8: Particles' p_T of the UE and of the low-p_T jets (from CKIN(3,108)). Due to the large statistics, the error bars are very small.

The match between the curves is not satisfying, as the "low-p_T jets" from CKIN(3,108) contain too many particles with a high p_T.
However, the "low-p_T jets" from CKIN(3,18) describe the UE particles in all p_T intervals of the leading jet quite well. Thus, it is now possible to approximate the hard scattering without UE by subtracting the constituents from the scaled "low-p_T jets" from hard events with UE. By dividing "hard+AMBT1"- ($sf \times$low-p_T jets) by "hard+AMBT1", weighting factors are derived:

$$w = \frac{"hard + AMBT1" - (sf \times low-p_T \ jets)}{"hard + AMBT1"} \quad , \tag{7.7}$$

with sf being the scaling factor.
With this statistically created probability distribution, it is possible to correct single events for UE contributions. For that purpose, the p_T of the jet constituents are weighted by the probability not to come from the UE. As can be seen in figure 7.9 (exemplarily for CKIN(3,18)) the probability for a particle to come from the UE is quite high for small p_T, whereas it diminishes for high values of p_T. A polynomial of the fifth order has been fitted to that curve. For each particle a weighting factor can be calculated by inserting the particle's momentum into the equation

$$weight = a_0 + a_1 \times p_T + a_2 \times (p_T)^2 + a_3 \times (p_T)^3 + a_4 \times (p_T)^4 + a_5 \times (p_T)^5 \ . \tag{7.8}$$

The prefactors of the fits are shown in table 7.4 for the different p_T intervals of the leading jets. Particles with $p_T < 30$ GeV are weighted by this function. At about 30 GeV the statistical fluctua-

---
[2]CKIN(3,p_T) stands for CKIN(3)=p_T.

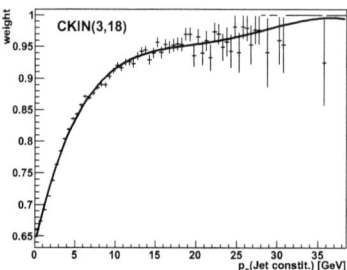

Figure 7.9: Weighting factors to correct for the UE

tions become very high. Moreover, the probability for a particle with more than 30 GeV is almost 1 to come from the hard scattering event and not from the UE. Hence, the weight is set to 1 in this area.

As a closure test, figures 7.10 show that the p_T distribution of the corrected (i.e. weighted) jet constituents from "hard+AMBT1" (labelled "hard + AMBT1$_{cor}$") is in good agreement with the curve from hard without UE (as a comparison, also the uncorrected jet constituents from "hard+AMBT1" have been included).

Thus the correction using a weight function obtained from the low-pT jets' particles allows to correct any significant bias in the jet-energy due to contributions from UE particles.

This method can also be used to correct the d_{23} values. For this reason, the flip-values have been calculated "by hand", using

$$d_{kB} = p_{Tk}^2 \quad \text{and} \quad d_{kl} = min(p_{Tk}^2, p_{Tl}^2) \times R_{kl}^2 \quad , \tag{7.9}$$

with the jet momenta p_{Tk}^2 and p_{Tl}^2 being the sum of the corrected (i.e. weighted) p_T of the jet constituents and

$$R_{kl}^2 = (\eta_k - \eta_l)^2 + (\Phi_k - \Phi_l)^2 \quad . \tag{7.10}$$

R_{kl} is not corrected in this case and is therefore identical with the original value, as it is assumed that the jet direction is not changed significantly by the UE.

The corrected d_{23} distributions ("hard + AMBT1$_{cor}$") are shown together with the uncorrected distribution in figures 7.11. The hard scattering without UE is also included into the plot.

The corrected d_{23} distributions match quite well with "hard", meaning that the low-p_T method can also be used to correct the flip-values for the contribution of the UE - especially at lower values of p_T of the leading jet. As the influence of the UE on the d_{23} distributions is small at higher values of CKIN(3), the "low-p_T method" shows no improvement in figures e) and f).

For a better illustration, the distribution of "hard + AMBT1$_{cor}$" is divided by "hard" in figures 7.12.

7.4. Low-p_T Method

p_T [GeV]	a_0 in 10^{-1}	a_1 in 10^{-2}	a_2 in 10^{-3}	a_3 in 10^{-4}	a_4 in 10^{-6}	a_5 in 10^{-8}
20-40	6.37 ± 1.91	5.59 ± 1.68	-3.95 ± 1.19	1.34 ± 0.40	-2.02 ± 0.61	1.02 ± 0.31
40-60	6.58 ± 3.03	8.40 ± 4.04	-8.35 ± 1.88	4.00 ± 0.62	-9.13 ± 1.86	7.96 ± 4.03
60-80	6.73 ± 3.01	8.83 ± 3.96	-9.23 ± 1.82	4.55 ± 0.60	-10.5 ± 1.75	9.28 ± 3.75
80-100	6.79 ± 3.05	9.20 ± 4.12	-10.1 ± 1.95	5.15 ± 0.65	-12.4 ± 1.97	11.2 ± 4.33
100-120	6.69 ± 3.38	10.8 ± 5.62	-13.8 ± 3.26	8.43 ± 1.35	-24.4 ± 5.00	26.8 ± 13.5
120-140	6.81 ± 3.25	10.3 ± 4.99	-12.7 ± 2.68	7.33 ± 1.02	-19.9 ± 3.51	20.5 ± 8.76

Table 7.4: Prefactors of the UE correction function

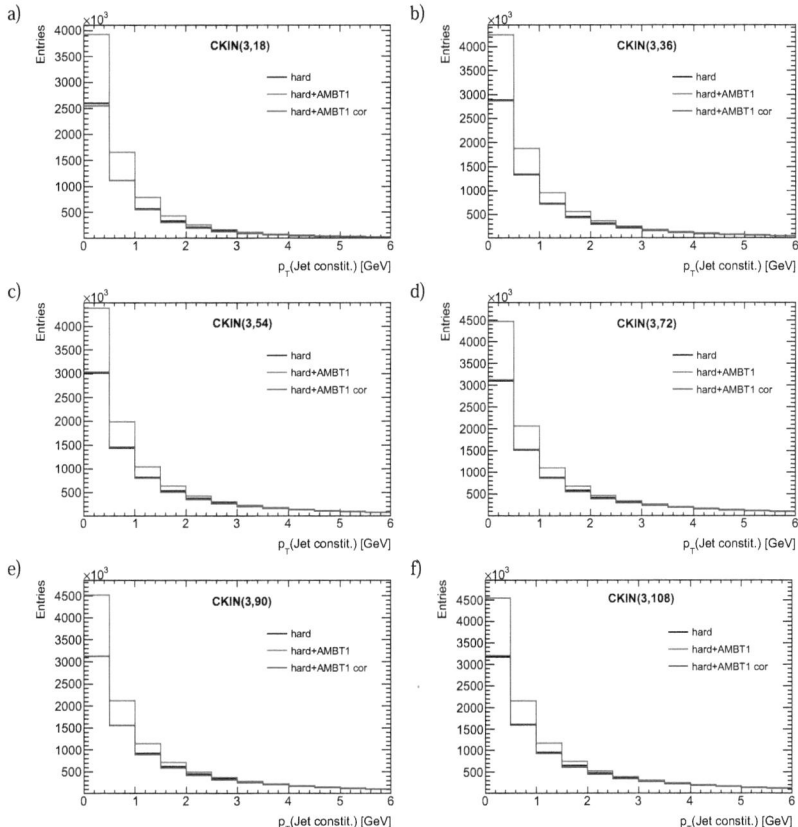

Figure 7.10: p_T distributions of the jet constituents before and after the correction compared to hard without UE. The black curve can hardly be seen because of the perfect match with the blue curve. Due to the large statistics, the error bars are very small.

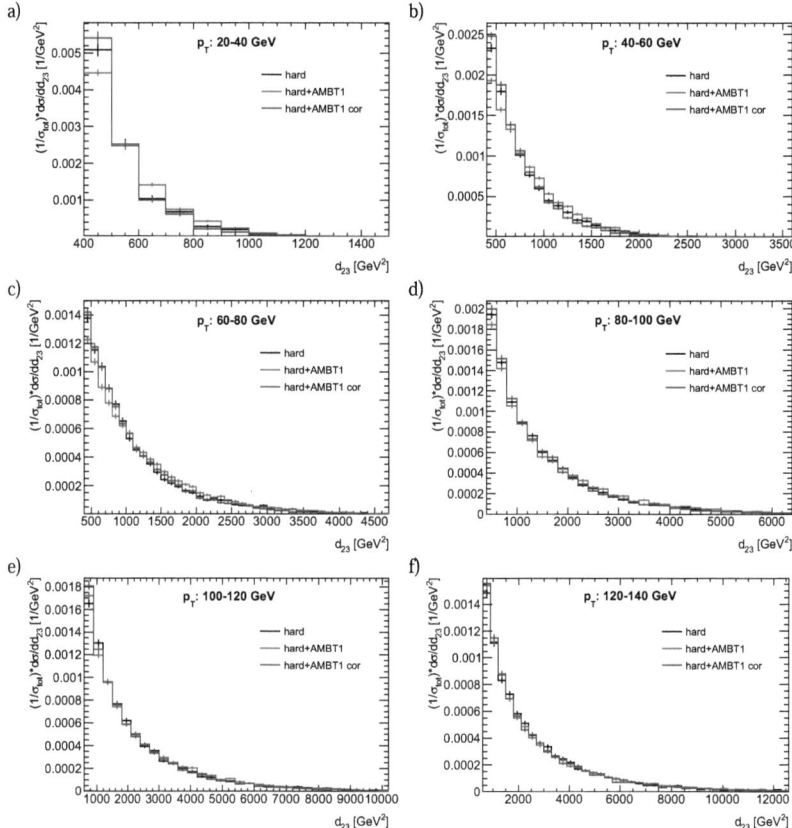

Figure 7.11: d_{23} distributions before and after the correction compared to hard without UE

p_T interval	Fit-value	Error
20-40 GeV	0.98	0.03
40-60 GeV	0.97	0.01
60-80 GeV	0.98	0.01
80-100 GeV	0.98	0.01
100-120 GeV	0.97	0.01
120-140 GeV	0.96	0.01

Table 7.5: Fit-values to the ratio of d_{23} distributions of "hard + AMBT1$_{cor}$" to hard only

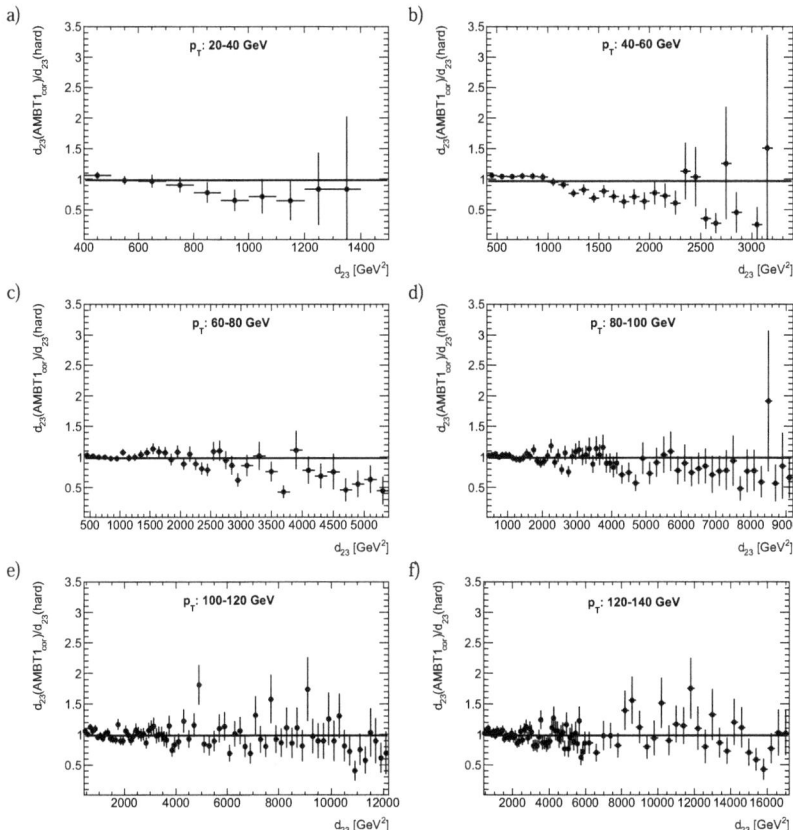

Figure 7.12: Ratio of d_{23} distributions of "hard + AMBT1$_{cor}$" to hard only. The black line represents a fit to the correction factors. The fit-values are shown in table 7.5.

Chapter 8

α_s-Fit and Systematic Uncertainties

In this chapter, α_s is determined at LO and NLO by fitting the calculations from NLOJET++ to data. Then, some systematic uncertainties are studied. Last but not least, α_s with its statistical and systematic uncertainties is presented and compared to the current world average.

8.1 α_s-Fit

α_s is determined by fitting the LO and NLO terms from NLOJET++ to data. In NLOJET++ the calculation of events is done separately for 2-parton-, 3-parton- and 4-parton-events. The leading order term of 2-parton-events is in the following called born$_2$, the next-to-leading order term nlo$_2$. Accordingly, born$_3$ and nlo$_3$ stand for the leading and next-to-leading order term of 3-parton-events and born$_4$ for the leading order term of 4-parton-events.
The total cross section is calculated in NLOJET++ via

$$\sigma_{tot} = \sigma_{born_2} + \sigma_{nlo_2} \ . \tag{8.1}$$

In this analysis, all events have been forced to a jet multiplicity of 3 in the final state. Events with bad jets have been excluded and jets with $p_T > 20$ GeV and $|\eta| < 2.6$ have been required. Moreover, a cut has been set at $d_{23} \geq 400$ GeV^2 in order to separate most of the 4-jet-events and to be safe from jet algorithm effects.
The total cross section for the studied 3-jet-events becomes

$$\sigma_{tot3}(Q) = \int_{400 \ GeV^2}^{\infty} \left(\frac{\mathrm{d}\sigma_{born2}(Q)}{\mathrm{d}d_{23}} + \frac{\mathrm{d}\sigma_{nlo2}(Q)}{\mathrm{d}d_{23}} \right) \mathrm{d}d_{23} = \sigma_{nlo_2}(Q) \ , \tag{8.2}$$

as $\frac{\mathrm{d}\sigma_{born2}(Q)}{\mathrm{d}d_{23}}$ does not contribute to 3-jet-events. Q is approximated by the p_T of the leading jet.
Before the fit of α_s is done, the distributions of d_{23} are normalized to one.
Therefore, real data are normalized by dividing $\frac{\Delta N(Q)}{\Delta d_{23}}$ by the number of entries with $d_{23} \geq 400$ GeV^2 in the according p_T interval of the leading jet ($p_{T,lj}$).

Accordingly,

$$\frac{\Delta born_3(d_{23}, Q)}{\Delta d_{23}} = \frac{1}{\sigma_{tot3}(Q)} \times \frac{\mathrm{d}\sigma_{born3}(Q)}{\mathrm{d}d_{23}} \tag{8.3}$$

and

$$\frac{\Delta born_4(d_{23}, Q)}{\Delta d_{23}} = \frac{1}{\sigma_{tot3}(Q)} \times \frac{\mathrm{d}\sigma_{born4}(Q)}{\mathrm{d}d_{23}} \ , \tag{8.4}$$

where $\sigma_{tot3}(Q)$ is treated as a constant value. As the events are forced to 3 jets in the final state, $\sigma_{tot3}(Q)$ always has the same number of entries, no matter if there have in fact been 3 jets, 4 jets or

even more jets in an event.
Therefore, α_s has been determined via the shape of the d_{23} distributions. The sensitivity of α_s on the shape of these distributions has been verified by changing the curve of real data: different shapes yielded different values of α_s.

For the χ^2-fit the α_s dependency has been made explicit for $\frac{d\sigma_{born3}(Q)}{dd_{23}}$ and $\frac{d\sigma_{born4}(Q)}{dd_{23}}$:

$$\frac{d\sigma_{born3}(Q)}{dd_{23}} = \frac{d\sigma_{born3}(Q)}{dd_{23}} \times \frac{\alpha_s(Q)^3}{\alpha_{s,int.(LO)}(Q)^3} \tag{8.5}$$

and

$$\frac{d\sigma_{born4}(Q)}{dd_{23}} = \frac{d\sigma_{born4}(Q)}{dd_{23}} \times \frac{\alpha_s(Q)^4}{\alpha_{s,int.(LO)}(Q)^4} \; . \tag{8.6}$$

The internal values of α_s cannot directly be accessed in NLOJET++ as they depend on Q. At LO, the running of $\alpha_{s,int.(LO)}$ is determined via [17]

$$\alpha_{s,int.(LO)} = \frac{1}{b_0 \times t} \times 2\pi \;\; \text{with} \;\; b_0 = 11 - \frac{2}{3} \times n_f \;\; \text{and} \;\; t = ln(Q/\Lambda) \; . \tag{8.7}$$

The number of active light flavors n_f is 5 in this analysis. Λ has a value of 0.2262 GeV [81]. For Q the center of the p_T intervals of the leading jets have been chosen.

In NLO the above formula becomes [17]

$$\alpha_{s,int.(NLO)} = \alpha_{s,int.(LO)} \times [1 - \frac{b_1}{b_0} \times \frac{\alpha_{s,int.(LO)}}{2\pi} \times ln(2 \times t)] \;\; \text{with} \;\; b_1 = 51 - \frac{19}{3} \times n_f \; . \tag{8.8}$$

Table 8.1 shows the resulting values of $\alpha_{s,int.}$ at LO and NLO in the according p_T intervals.

p_T Interval	$\alpha_{s,int.(LO)}$	$\alpha_{s,int.(NLO)}$	d_{23} ($I \leq 20\%$) in GeV^2	d_{23} ($I \leq 12.5\%$) in GeV^2
20-40 GeV	0.167681	0.141954	400	600
40-60 GeV	0.151814	0.129806	600	900
60-80 GeV	0.142907	0.122910	700	1200
80-100 GeV	0.136907	0.118231	900	1500
100-120 GeV	0.132467	0.114750	1200	1900
120-140 GeV	0.128984	0.112009	1400	2300

Table 8.1: Internal values of α_s at LO and NLO as well as d_{23} values with an impurity (of 4-jet-events) of 20% and 12.5%

8.1.1 LO α_s

In this section, α_s is determined at leading order. For this reason, a χ^2-fit has been applied:

$$\chi^2 = \sum_{d_{23}(I)}^{\infty} \left[\left(\frac{1}{N(Q)} \times \frac{\Delta N(Q)}{\Delta d_{23}} - \frac{\Delta born_3(d_{23}, Q)}{\Delta d_{23}} \times \frac{\alpha_s(Q)^3}{\alpha_{s,int.(LO)}(Q)^3} \right) \Big/ \sigma_{Data,stat.} \right]^2 , \tag{8.9}$$

8.1. α_s-Fit

with $d_{23}(I)$ representing the value of d_{23} where the impurity (I) due to 4-jet-events is at most 20% and 12.5% respectively (see chapter 6.3.3). The according values are shown in table 8.1.

As the statistical errors from NLOJET++ are very small, only the statistical errors from data ($\sigma_{Data,stat.}$), i.e. the square root of the numbers of entries in a bin divided by the number of analyzed events in the according $p_{T,IJ}$ interval, have been taken into account.

In figure 8.1 the running of α_s depending on $p_{T,IJ}$ at LO is shown for the jet triggers L1_J15 and L1_J30 with an impurity of 4-jet-events of 20%. Data have been adjusted for the jet-energy-scale (see chapter 7.1) and the calculations from NLOJET++ have been corrected for hadronization effects (see chapter 7.2) and the influence of the Underlying Event (see chapter 7.3).

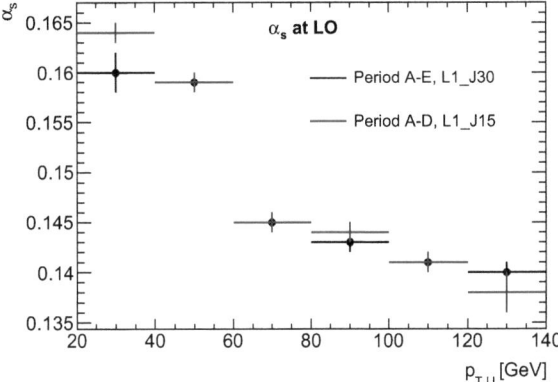

Figure 8.1: Running of α_s at LO using trigger L1_J15 (run periods A-D) and trigger L1_J30 (run periods A-E)

In this figure, only the statistical errors ($\Delta\alpha_s^{stat.}$) are shown (the systematic uncertainties are studied in chapter 8.2).
For both triggers solely un-prescaled run periods have been analyzed. Therefore, run periods A-D have been used for trigger L1_J15 and run periods A-E for trigger L1_J30, resulting in larger statistical errors when using trigger L1_J15. However, at the $p_{T,IJ}$ interval between 20 GeV and 40 GeV the statistical error of α_s using trigger L1_J30 is larger: Due to the minimum p_T of 30 GeV several events are lost in this bin, resulting in less statistics. Apart from the first bin, the values of α_s agree within the statistical fluctuations.
The values of α_s at LO using trigger L1_J30 are shown together with the statistical errors and systematic uncertainties in table 8.7 (see chapter 8.3).

8.1.2 NLO α_s

The value of α_s is in the following determined at next-to-leading order.
As shown in chapter 6.2.2, nlo$_2$ is almost identical with born$_3$, besides smaller differences due to virtual corrections at very small values of d_{23}. Besides these virtual corrections, the calculation in NLOJET++ is the same for nlo$_n$ and born$_{n+1}$ (with n being the number of partons).
In this analysis, only values of at least $d_{23} \geq 400$ GeV^2 have been studied. In this region, the virtual corrections are insignificant. Therefore, in order to avoid double counting, only born$_3$ (which already

includes nlo$_2$) and born$_4$ (which already includes nlo$_3$) have been considered for the α_s-fit.
The formula of the χ^2-fit at NLO becomes:

$$\chi^2 = \sum_{d_{23}(I)}^{\infty} \left[\left(\frac{1}{N(Q)} \times \frac{\Delta N(Q)}{\Delta d_{23}} - \left(\frac{\Delta born_3(d_{23}, Q)}{\Delta d_{23}} \times \frac{\alpha_s(Q)^3}{\alpha_{s,int.(LO)}(Q)^3} \right) \right. \right.$$
$$\left. \left. - \left(\frac{\Delta born_4(d_{23}, Q)}{\Delta d_{23}} \times \frac{\alpha_s(Q)^4}{\alpha_{s,int.(LO)}(Q)^4} \right) \right) \bigg/ \sigma_{Data,stat.} \right]^2 . \quad (8.10)$$

The same cuts and corrections have been applied as in chapter 8.1.1. The results of the NLO determination of α_s using the jet triggers L1_J15 and L1_J30 are shown in figure 8.2.

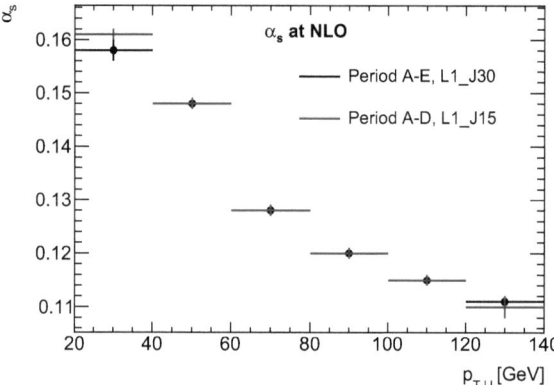

Figure 8.2: Running of α_s at NLO using trigger L1_J15 (run periods A-D) and trigger L1_J30 (run periods A-E)

α_s determined with data satisfying the L1_J15 trigger condition has again larger statistical errors than data from trigger L1_J30, due to the missing run period E for trigger L1_J15[1].
The values of α_s applying different triggers agree within the statistical fluctuations. As the statistics are higher for trigger L1_J30, only this trigger has been used in the following to study systematic uncertainties.
α_s at NLO using trigger L1_J30 is presented with its statistical and systematic uncertainty in table 8.8 (see chapter 8.3).

---

[1]The lager statistical error of L1_J30 in the first bin is due to the trigger criterion, demanding a minimum p_T of 30 GeV. Therefore, several events not passing the trigger are lost.

8.2 Systematic Uncertainties

In this chapter, some systematic uncertainties are studied. First of all, the uncertainty of the jet-energy-scale is determined. Then, the impurity due to 4-jet-events is studied, followed by an investigation of the renormalization scale and the PDF uncertainty, comparing different PDF sets. The subsequent section describes the uncertainty of the hadronization, comparing the results from PYTHIA and HERWIG. Afterwards, the uncertainty of the Underlying Event is measured by comparing different UE models.

8.2.1 JES Uncertainty

In chapter 7.1 the jet-energy-scale has been corrected to compensate detector effects. For the anti-k_T algorithm the JES currently has an uncertainty of around 5% [52]. A conservative approach is therefore to take this uncertainty twice to estimate the JES uncertainty of the k_T jet algorithm in the exclusive mode. The correction factors for the JES have such been increased by 10%. As this systematic uncertainty is considered to be symmetric [89], a decrease of the JES of 10% has also been studied.

Figures 8.3 show the α_s distribution within a variation of the JES at LO and NLO.

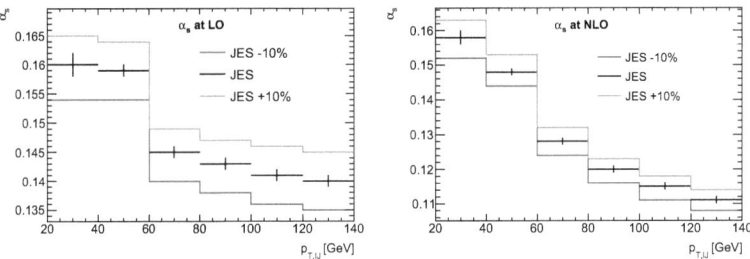

Figure 8.3: JES uncertainty of α_s at LO (left) and NLO (right). The correction for the JES has been varied by 10%.

The JES uncertainty in the first two bins is potentially larger as shown in the figures, because of bin migrations from $p_{T,1J} < 20$ GeV due to the shift of the energy-scale (see chapter 7.1).
The values of the JES uncertainty ($\Delta \alpha_s^{sys.}(JES)$) are shown in table 8.2. This uncertainty has a maximum of around 3.8%.

The correction factors

$$H_{JES}^j(d_{23}) = \frac{d_{23,truth}^j}{d_{23,reco}^j} \qquad (8.11)$$

from chapter 7.1 have in addition statistical errors, which contribute as systematic uncertainties to the α_s determination. To study this uncertainty, the statistical errors of $H_{JES}^j(d_{23})$ have been quadratically added to the statistical errors of measured data ($\sigma_{Data,stat.}$). α_s has then been fitted by varying the JES correction according to the total statistical errors ($\sigma_{stat.}$).
The results are shown in figures 8.4.
In order to evaluate the systematic uncertainty due to the statistical errors of the JES ($\Delta \alpha_s^{sys.}(JES, stat.)$), the statistical errors of α_s (due to data) have to be subtracted quadratically to avoid double counting of the statistical errors of data:

$$\Delta \alpha_s^{sys.}(JES, stat.) = \sqrt{[\Delta \alpha_s^{stat.}(JES + Data)]^2 - [\Delta \alpha_s^{stat.}]^2} \quad . \qquad (8.12)$$

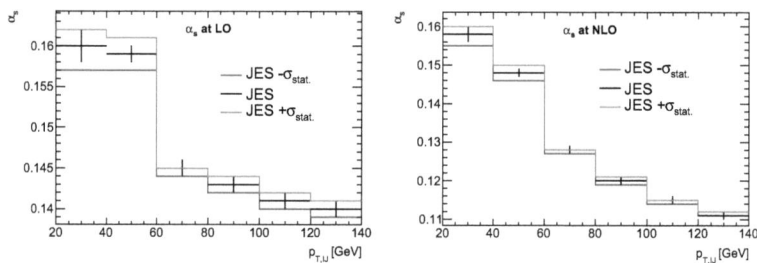

Figure 8.4: Statistical uncertainty of JES and data at LO (left) and NLO (right). The correction for the JES has been varied by the total statistical errors ($\sigma_{stat.}$).

The values of $\Delta\alpha_s^{sys.}(JES, stat.)$ are presented in table 8.2.

When summarizing all systematic uncertainties in tables 8.7 and 8.8 (see chapter 8.3) always the largest value of the according uncertainty has been taken.

8.2. Systematic Uncertainties

LO

p_T Interval	α_s	$\Delta\alpha_s^{sys.}(JES)$ [JES -10%]	$\Delta\alpha_s^{sys.}(JES)$ [JES +10%]	$\Delta\alpha_s^{sys.}(JES, stat.)$ [$-\sigma_{stat.}$]	$\Delta\alpha_s^{sys.}(JES, stat.)$ [$+\sigma_{stat.}$]
20-40 GeV	0.160	0.006	0.005	0.002	<0.001
40-60 GeV	0.159	0.005	0.005	0.002	0.002
60-80 GeV	0.145	0.005	0.004	<0.001	<0.001
80-100 GeV	0.143	0.005	0.004	<0.001	<0.001
100-120 GeV	0.141	0.005	0.005	<0.001	<0.001
120-140 GeV	0.140	0.005	0.005	<0.001	<0.001

NLO

p_T Interval	α_s	$\Delta\alpha_s^{sys.}(JES)$ [JES -10%]	$\Delta\alpha_s^{sys.}(JES)$ [JES +10%]	$\Delta\alpha_s^{sys.}(JES, stat.)$ [$-\sigma_{stat.}$]	$\Delta\alpha_s^{sys.}(JES, stat.)$ [$+\sigma_{stat.}$]
20-40 GeV	0.158	0.006	0.005	0.002	<0.001
40-60 GeV	0.148	0.004	0.005	0.002	0.002
60-80 GeV	0.128	0.004	0.004	<0.001	<0.001
80-100 GeV	0.120	0.004	0.003	<0.001	<0.001
100-120 GeV	0.115	0.004	0.003	<0.001	<0.001
120-140 GeV	0.111	0.003	0.003	<0.001	<0.001

Table 8.2: Systematic uncertainty of α_s due to the JES. The JES correction has been changed by $\pm 10\%$. In addition, the uncertainty $\Delta\alpha_s^{sys.}(JES, stat.)$ is presented, which has been determined by varying the JES correction by $\pm\sigma_{stat.}$ (i.e. the total statistical error) and quadratically subtracting the statistical error of α_s from the resulting α_s uncertainty.

8.2.2 Impurity due to 4-Jet-Events

In chapter 6.3.3 the impurity of the 3-jet-rate due to 4-jet-events has been studied and cut values of d_{23} have been determined for impurities (of d_{34}) of 20% and 12.5%. These cut values are shown in table 8.1.
In figures 8.5 α_s is shown for the different impurities.

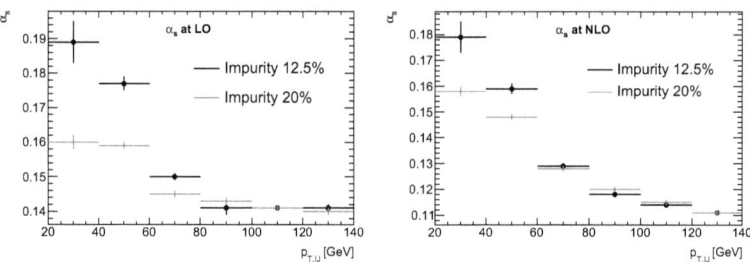

Figure 8.5: α_s at LO (left) and NLO (right), determined with an impurity due to 4-jet-events of 20% and 12.5%

For large values of $p_{T,U}$ the curves agree within the statistical fluctuations, whereas the impact of 4-jet-events become important at small values of $p_{T,U}$. With a smaller impurity due to 4-jet-events (i.e. higher cut values on d_{23}), the statistics decrease. Thus, the statistical errors are larger for an impurity of 12.5% than for an impurity of 20%. Therefore, always an impurity of 20% has been studied in the following.
As a systematic uncertainty of the impurity due to 4-jet-events, the difference of the two distributions has been taken. The uncertainty is quite large in first bin with 18.1% at LO and 13.3% at NLO and decreases to less than 1% for high values of $p_{T,U}$. The according uncertainties ($\Delta\alpha_s^{sys.}(4jet)$) are shown in tables 8.7 and 8.8 (see chapter 8.3).

8.2.3 Renormalization Scale Uncertainty

The value of α_s depends on the energy. In order to handle singularities of Feynman diagrams, a renormalization scale is needed in theory (for more details see e.g. [90]). The calculations with NLOJET++ have been done at a renormalization scale factor (RSF) of 1. To study the theoretical uncertainty of the renormalization scale, the calculations have also been done at renormalization scale factors of 0.5 and 2.0, respectively.
α_s has then been determined for the different values (see figures 8.6).

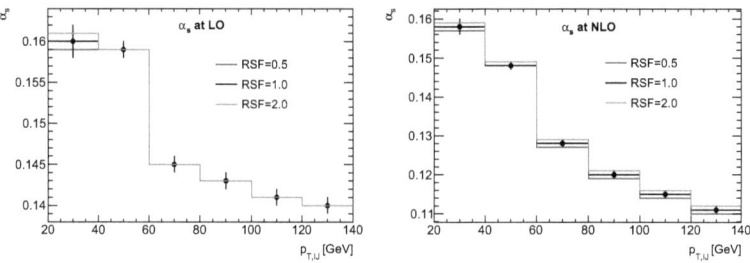

Figure 8.6: α_s at LO (left) and NLO (right), determined for different renormalization scale factors (RSF)

8.2. Systematic Uncertainties

Table 8.3 shows the according uncertainties.

p_T Interval	LO		NLO	
	$\Delta\alpha_s^{sys.}(RS)$ [RSF=0.5]	$\Delta\alpha_s^{sys.}(RS)$ [RSF=2.0]	$\Delta\alpha_s^{sys.}(RS)$ [RSF=0.5]	$\Delta\alpha_s^{sys.}(RS)$ [RSF=2.0]
20-40 GeV	0.001	0.001	0.001	0.001
40-60 GeV	<0.001	<0.001	<0.001	0.001
60-80 GeV	<0.001	<0.001	0.001	0.001
80-100 GeV	<0.001	<0.001	0.001	0.001
100-120 GeV	<0.001	<0.001	0.001	0.001
120-140 GeV	<0.001	<0.001	0.001	0.001

Table 8.3: Systematic uncertainty of α_s due to the renormalization scale. Scale factors of 0.5 and 2.0 have been studied.

The uncertainty due to different factors of the renormalization scale is always smaller than 1%. The maximum value of the theoretical uncertainty due to the renormalization scale has been included as a systematic uncertainty ($\Delta\alpha_s^{sys.}(RS)$) into tables 8.7 and 8.8 (see chapter 8.3).

8.2.4 PDF Uncertainty

The PDF CTEQ66M has been used in this analysis to calculate the cross sections with NLOJET++ at LO and NLO. The influence of different PDF sets has been studied with PYTHIA on parton level, as the simulation with PYTHIA is much faster than with NLOJET++.

Figures 8.7 compare d_{23} distributions to each other using various different PDF sets.

The NLO PDF CTEQ66M has been chosen, as it is the default PDF set in NLOJET++ and also provides corrections for an additional virtual parton. It has been compared to CTEQ5L [22], which is the standard PDF set in PYTHIA, providing a LO PDF. This is the oldest of the compared PDF sets. A next-to-leading-log-approximation (NLLA) is provided by CTEQ5M1. Hence, corrections of higher orders to a dominant term are considered. However, the NLLA does not provide high accuracy of the higher order terms [2]. The LO CTEQ6L with a NLO α_s has also been compared. Finally, the newest PDF set MSTW2008 is also included into the figure.

The d_{23} distributions show just small differences due to the different PDF sets.

As α_s has been determined in this analysis, the systematic uncertainty of the PDF is measured by using CTEQ66M for different values of α_s.

Figures 8.8 compare the d_{23} distributions simulated with CTEQ66M with a PDF α_s of 0.117, 0.118 and 0.119. The range corresponds to one standard deviation about the world average.

To determine the systematic uncertainty, the calculations from NLOJET++ have been corrected for the different PDF α_s values. The results of the α_s-fits are shown in figures 8.9.

The PDF uncertainty is not very large with a maximum of around 1.9% at LO and 1.4% at NLO. Its values are shown for CTEQ66M with a PDF α_s of 0.117 and 0.119 in table 8.4.

In tables 8.7 and 8.8 (see chapter 8.3) the largest values of the presented PDF uncertainties ($\Delta\alpha_s^{sys.}(PDF)$) have been taken.

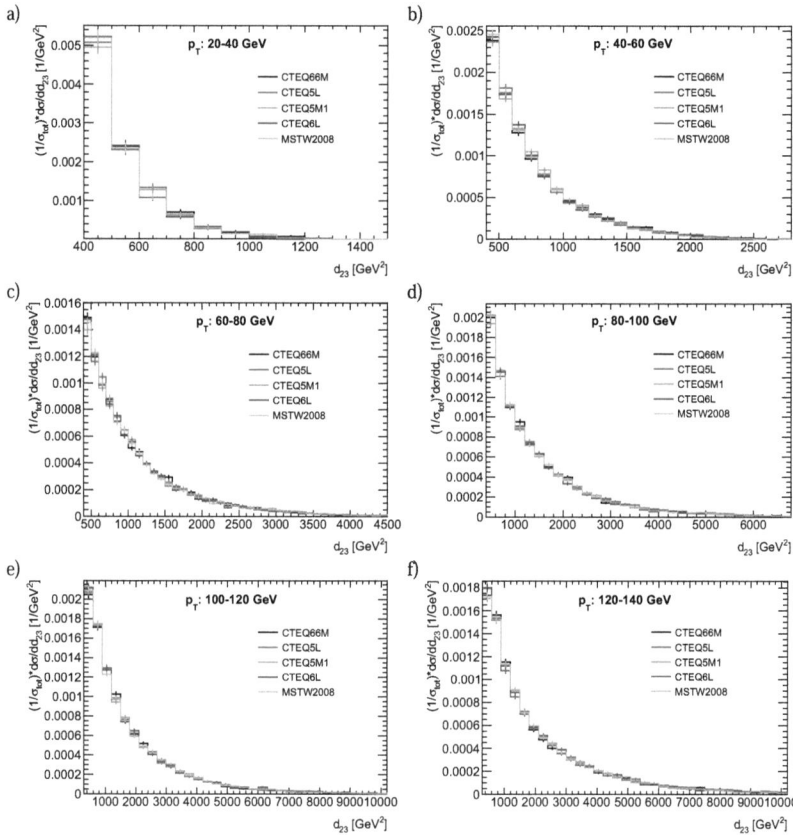

Figure 8.7: Influence of different PDFs on the d_{23} distribution. The p_T of the leading jet is inside the according interval.

p_T Interval	LO		NLO	
	$\Delta\alpha_s^{sys.}(PDF)$ [PDF $\alpha_s = 0.017$]	$\Delta\alpha_s^{sys.}(PDF)$ [PDF $\alpha_s = 0.019$]	$\Delta\alpha_s^{sys.}(PDF)$ [PDF $\alpha_s = 0.017$]	$\Delta\alpha_s^{sys.}(PDF)$ [PDF $\alpha_s = 0.019$]
20-40 GeV	0.001	0.001	<0.001	0.001
40-60 GeV	0.002	0.003	0.002	0.002
60-80 GeV	0.001	<0.001	<0.001	<0.001
80-100 GeV	0.001	<0.001	0.001	<0.001
100-120 GeV	0.001	<0.001	0.001	<0.001
120-140 GeV	0.001	0.001	<0.001	<0.001

Table 8.4: Systematic uncertainty of α_s due to PDF, using CTEQ66M with a PDF α_s of 0.117 and 0.119

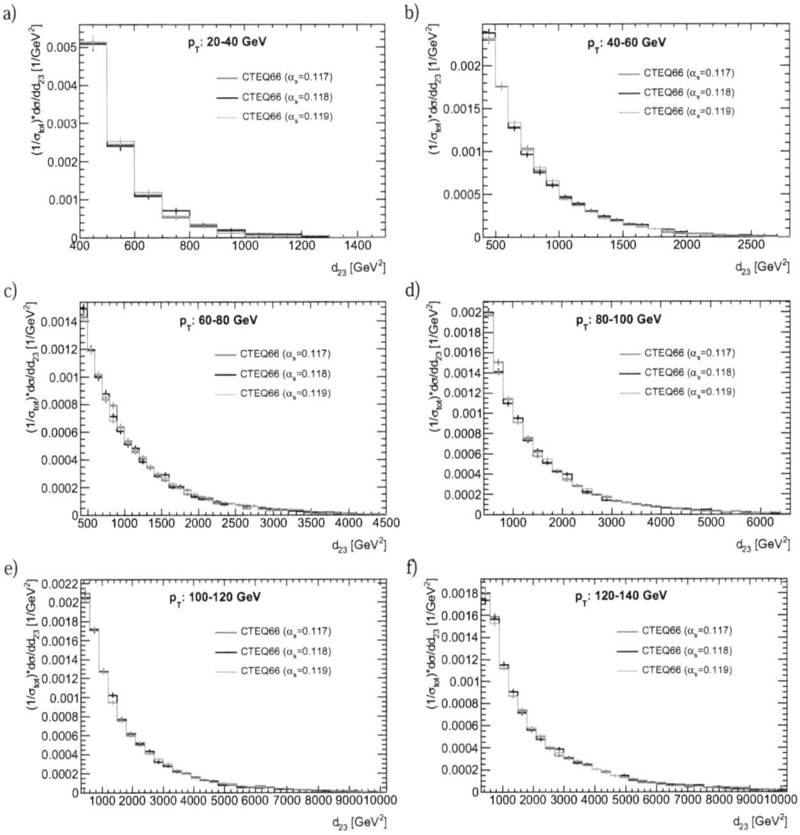

8.2. Systematic Uncertainties

Figure 8.8: d_{23} distributions, comparing CTEQ66M with different values of α_s. The p_T of the leading jet is inside the according interval.

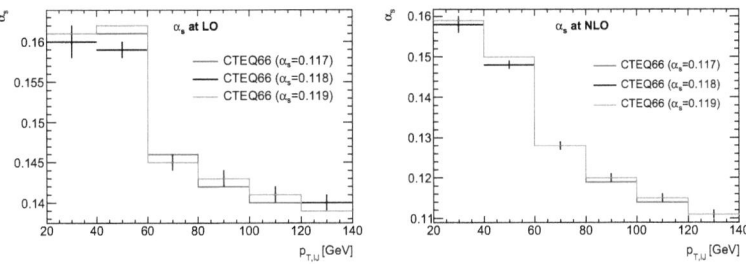

Figure 8.9: Systematic uncertainty of α_s at LO (left) and NLO (right) due to different PDF sets. CTEQ66M has been used with $\alpha_s = 0.117$, $\alpha_s = 0.118$ and $\alpha_s = 0.119$.

8.2.5 Hadronization Uncertainty

In order to take the hadronization into account, PYTHIA has been used to simulate events on parton and on hadron level. The d_{23} distributions of the hadron and parton level simulations have been divided to get correction factors (see chapter 7.2). The calculations of NLOJET++ have then been corrected bin-by-bin before being used in the α_s-fit.
For systematic studies, the program HERWIG has been used, as it also allows the simulation on parton as well as on hadron level and additionally uses a different hadronization model (see chapter 2.3). The correction factors from PYTHIA are compared to HERWIG in figure 8.10.

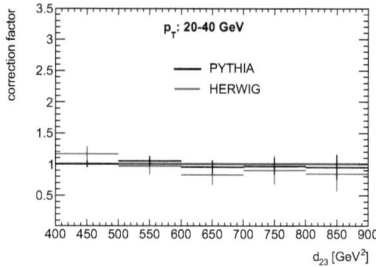

Figure 8.10: Hadronization correction factors from PYTHIA and HERWIG. The fit to the correction factors from HERWIG yields a value of 1.001 ± 0.007.

As the impact of the hadronization is large at small values of p_T of the leading jet and decreases with higher energies, only the p_T interval from 20 GeV to 40 GeV has been investigated.

The fit to the correction factors from HERWIG yields 1.001 ± 0.007, whereas the fit to the correction factors from PYTHIA gave a value of 0.995 ± 0.004 (see table 7.1 in chapter 7.2).
Dividing 1.001 by 0.995 yields a factor of 1.006. As a conservative approach, this was rounded to 1.01 and α_s determined after scaling the NLOJET++ calculations by this factor. α_s at LO and NLO considering the hadronization uncertainties is displayed in figures 8.11.

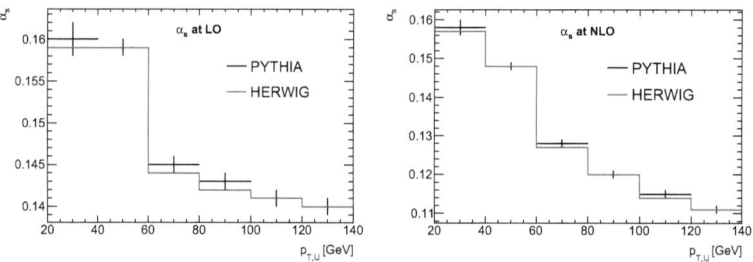

Figure 8.11: Hadronization uncertainty at LO (left) and NLO (right)

The hadronization uncertainty ($\Delta\alpha_s^{sys.}(hadr.)$), which is always smaller than 1%, is displayed in table 8.5.
The systematic uncertainty due to the statistical errors of the correction factors (see chapter 7.2)

$$H_{hadr}^j(d_{23}) = \frac{d_{23,hadr}^j}{d_{23,part}^j} \qquad (8.13)$$

8.2. Systematic Uncertainties

have also been studied.
The α_s distributions accounting for the statistical errors of $H^i_{hadr}(d_{23})$ (in the following called $\sigma_{hadr.}$) are presented in figures 8.12.

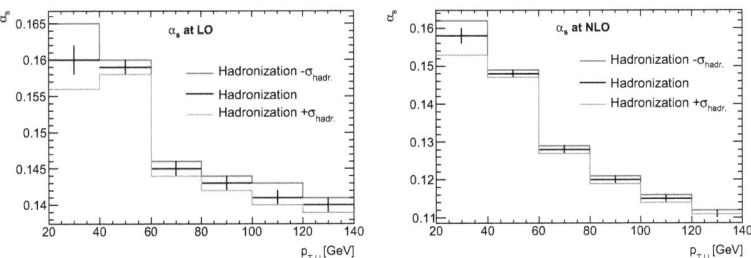

Figure 8.12: Systematic uncertainty due to statistical errors of the hadronization correction at LO (left) and NLO (right)

The uncertainties are quite large in the first bin (around 3%) and have then values of around 1%. The values of the systematic uncertainties of α_s due to the statistical fluctuations of the correction factors $(\Delta\alpha_s^{sys.}(hadr., stat.))$ are shown in table 8.5.

p_T Interval	α_s	$\Delta\alpha_s^{sys.}(hadr.)$	$\Delta\alpha_s^{sys.}(hadr., stat.)$ [$-\sigma_{hadr.}$]	$\Delta\alpha_s^{sys.}(hadr., stat.)$ [$+\sigma_{hadr.}$]
		LO		
20-40 GeV	0.160	0.001	0.005	0.004
40-60 GeV	0.159	<0.001	0.001	0.001
60-80 GeV	0.145	0.001	0.001	0.001
80-100 GeV	0.143	0.001	0.001	0.001
100-120 GeV	0.141	<0.001	0.002	0.001
120-140 GeV	0.140	<0.001	0.001	0.001
		NLO		
20-40 GeV	0.158	0.001	0.004	0.005
40-60 GeV	0.148	<0.001	0.001	0.001
60-80 GeV	0.128	0.001	0.001	0.001
80-100 GeV	0.120	<0.001	0.001	0.001
100-120 GeV	0.115	0.001	0.001	0.001
120-140 GeV	0.111	<0.001	0.001	<0.001

Table 8.5: Systematic uncertainty due to the hadronization and due to the statistical errors of the hadronization correction

Always the largest value of the presented hadronization uncertainties has been included into tables 8.7 and 8.8 in chapter 8.3.

8.2.6 Underlying Event Uncertainty

The systematic uncertainty of the Underlying Event (UE) has been determined by correcting the theory predictions from NLOJET++ for the impact of the UE, using different UE models (see chapter 7.3). The UE has been simulated with PYTHIA, applying the UE tunes AMBT1, PERUGIA10 and ATLAS MC09c (see chapter 5.2.2).
The distribution of α_s at LO and NLO where the effects of each UE model considered have been absorbed in the NLOJET++ prediction is shown in figures 8.13.

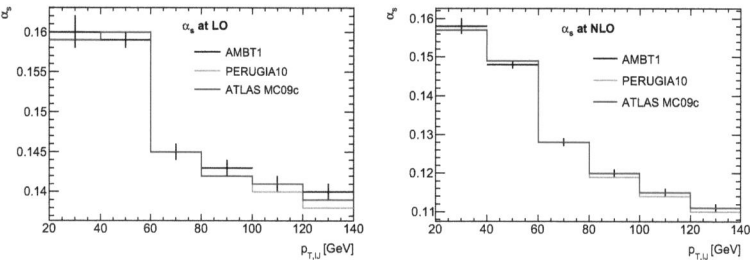

Figure 8.13: Underlying Event uncertainty at LO (left) and NLO (right)

The distributions show only small differences of around 1%. The values of the UE uncertainty are shown in table 8.6.

	LO		NLO	
p_T Interval	PERUGIA10	ATLAS MC09	PERUGIA10	ATLAS MC09
20-40 GeV	0.001	0.001	0.001	0.001
40-60 GeV	0.001	0.001	0.001	0.001
60-80 GeV	<0.001	<0.001	<0.001	<0.001
80-100 GeV	0.001	0.001	0.001	<0.001
100-120 GeV	0.001	<0.001	0.001	<0.001
120-140 GeV	0.002	0.001	0.001	<0.001

Table 8.6: Systematic uncertainty of α_s due to the UE, using the UE models PERUGIA10 and ATLAS MC09

For the total systematic uncertainty (see tables 8.7 and 8.8 in the next chapter), the largest uncertainty of the UE tunes ($\Delta\alpha_s^{sys.}(UE)$) has been chosen.

The same events have been simulated as for the hadronization corrections. Therefore, the statistical error of the UE correction has already been considered in the above chapter.

8.3 Final Results

In this section, the final results are presented and compared to the world average.
The above sections determined various systematic uncertainties. A summary also including the statistical uncertainties is given in table 8.7 at LO and in table 8.8 at NLO.

LO

$p_{T,1J}$ Interval	α_s	$\Delta\alpha_s^{stat.}$	$\Delta\alpha_s^{sys.}(JES)$	$\Delta\alpha_s^{sys.}(JES, stat.)$	$\Delta\alpha_s^{sys.}(4\,jet)$	$\Delta\alpha_s^{sys.}(RS)$
20-40 GeV	0.160	0.002	0.006	0.002	0.029	0.001
40-60 GeV	0.159	0.001	0.005	0.002	0.018	<0.001
60-80 GeV	0.145	0.001	0.005	<0.001	0.005	<0.001
80-100 GeV	0.143	0.001	0.005	<0.001	0.002	<0.001
100-120 GeV	0.141	0.001	0.005	<0.001	<0.001	<0.001
120-140 GeV	0.140	0.001	0.005	<0.001	0.001	<0.001

$p_{T,1J}$ Interval	$\Delta\alpha_s^{sys.}(PDF)$	$\Delta\alpha_s^{sys.}(hadr.)$	$\Delta\alpha_s^{sys.}(hadr., stat.)$	$\Delta\alpha_s^{sys.}(UE)$	$\Delta\alpha_s^{sys.}(total)$
20-40 GeV	0.001	0.001	0.005	0.001	0.030
40-60 GeV	0.003	<0.001	0.001	0.001	0.019
60-80 GeV	0.001	0.001	0.001	<0.001	0.007
80-100 GeV	0.001	0.001	0.001	0.001	0.006
100-120 GeV	0.001	<0.001	0.002	0.001	0.006
120-140 GeV	0.001	<0.001	0.001	0.002	0.006

Table 8.7: Statistical ($\Delta\alpha_s^{stat.}$) and systematic uncertainties of α_s at LO. The uncertainty of the JES ($\Delta\alpha_s^{sys.}(JES)$), the systematic uncertainty due to the statistical errors of the JES ($\Delta\alpha_s^{sys.}(JES, stat.)$), the uncertainty due to 4-jet-events ($\Delta\alpha_s^{sys.}(4\,jet)$), the uncertainty of the renormalization scale ($\Delta\alpha_s^{sys.}(RS)$), the uncertainty of the PDF ($\Delta\alpha_s^{sys.}(PDF)$), the uncertainty of the hadronization ($\Delta\alpha_s^{sys.}(hadr.)$), the systematic uncertainty due to the statistical errors of the hadronization correction ($\Delta\alpha_s^{sys.}(hadr., stat.)$) and the uncertainty of the UE ($\Delta\alpha_s^{sys.}(UE)$) have been added to the total systematic uncertainty ($\Delta\alpha_s^{sys.}(total)$), where values <0.001 have been handled as 0.001. In this table, always the largest uncertainty has been included.

In the first two $p_{T,1J}$ intervals, bin-migration effects from $p_{T,1J} < 20$ GeV have potentially larger effects on the estimation of the systematic uncertainties. These effects are insignificant at higher p_T intervals.

α_s at NLO has been compared to the theory curve defined at the value of the world average of $\alpha_s(M_Z) = 0.1184 \pm 0.0007$ (see figure 8.14).

In green, the total systematic uncertainty is shown. The data points are in good agreement with the theory curve - especially in the region $p_{T,1J} > 60$ GeV.
At the mass of the Z boson, α_s has been determined to $\alpha_s(M_Z) = 0.120 \pm 0.001(stat.) \pm 0.005(syst.)$, which is in good agreement with the current world average.

NLO

$p_{T,lJ}$ Interval	α_s	$\Delta\alpha_s^{stat.}$	$\Delta\alpha_s^{sys.}(JES)$	$\Delta\alpha_s^{sys.}(JES,stat.)$	$\Delta\alpha_s^{sys.}(4jet)$	$\Delta\alpha_s^{sys.}(RS)$
20-40 GeV	0.158	0.002	0.006	0.002	0.021	0.001
40-60 GeV	0.148	0.001	0.005	0.002	0.011	0.001
60-80 GeV	0.128	0.001	0.004	<0.001	0.001	0.001
80-100 GeV	0.120	0.001	0.004	<0.001	0.002	0.001
100-120 GeV	0.115	0.001	0.004	<0.001	0.001	0.001
120-140 GeV	0.111	0.001	0.003	<0.001	<0.001	0.001

$p_{T,lJ}$ Interval	$\Delta\alpha_s^{sys.}(PDF)$	$\Delta\alpha_s^{sys.}(hadr.)$	$\Delta\alpha_s^{sys.}(hadr.,stat.)$	$\Delta\alpha_s^{sys.}(UE)$	$\Delta\alpha_s^{sys.}(total)$
20-40 GeV	0.001	0.001	0.005	0.001	0.023
40-60 GeV	0.002	<0.001	0.001	0.001	0.013
60-80 GeV	<0.001	0.001	0.001	<0.001	0.005
80-100 GeV	0.001	<0.001	0.001	0.001	0.005
100-120 GeV	0.001	0.001	0.001	0.001	0.005
120-140 GeV	<0.001	<0.001	0.001	0.001	0.004

Table 8.8: Statistical ($\Delta\alpha_s^{stat.}$) and systematic uncertainties of α_s at NLO. The uncertainty of the JES ($\Delta\alpha_s^{sys.}(JES)$), the systematic uncertainty due to the statistical errors of the JES ($\Delta\alpha_s^{sys.}(JES,stat.)$), the uncertainty due to 4-jet-events ($\Delta\alpha_s^{sys.}(4jet)$), the uncertainty of the renormalization scale ($\Delta\alpha_s^{sys.}(RS)$), the uncertainty of the PDF ($\Delta\alpha_s^{sys.}(PDF)$), the uncertainty of the hadronization ($\Delta\alpha_s^{sys.}(hadr.)$), the systematic uncertainty due to the statistical errors of the hadronization correction ($\Delta\alpha_s^{sys.}(hadr.,stat.)$) and the uncertainty of the UE ($\Delta\alpha_s^{sys.}(UE)$) have been added to the total systematic uncertainty ($\Delta\alpha_s^{sys.}(total)$), where values <0.001 have been handled as 0.001. In this table, always the largest uncertainty has been included.

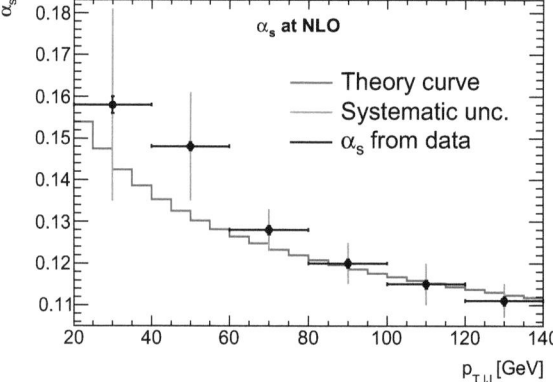

Figure 8.14: α_s compared to theory curve defined at the value of the world average of $\alpha_s(M_Z) = 0.1184 \pm 0.0007$. In green, the total systematic uncertainty of α_s is shown.

Chapter 9
Summary

The ATLAS detector at LHC started to record data from pp collisions at a center-of-mass-energy of 7 TeV on March 30th 2010. When starting a new experiment, first of all the detector has to be understood and it has to be shown that the experiment works well, reproduces the results from former colliders and is consistent with the theoretical extrapolation to the high collision energies of the LHC. As a test of QCD, this thesis has determined the strong coupling constant α_s via the ratio of 3-jet-events to 2-jet-events with an integrated luminosity of 700 nb^{-1}. As the entries of jet-rates are correlated to each other, it is preferable to analyze differential jet-rates instead. The differential 2-jet-rate is measured via the jet-flip-values (d_{23}), describing the transition from 3 to 2 reconstructed jets.

The transition parameter from 3 to 2 jets from the Durham jet algorithm has already been used for the α_s determination in former collider experiments, especially at e^+e^- colliders. This flip-parameter is equivalent to the measurement of the ratio of trijet to dijet events. In this way, the theoretical uncertainties can be reduced as many of them almost cancel out.

Analogously, in this analysis, the k_T algorithm in the exclusive reconstruction mode has been used for the α_s determination. It is based on the Durham jet algorithm and therefore also allows access to the flip-values from 3 to 2 reconstructed jets, with the algorithm being forced to find 3 jets in the final state. These flip-values are not very sensitive to the jet-energy-scale and hence allowed the measurement of α_s at an early stage of the experiment.

Unfortunately, the k_T algorithm in the exclusive mode is not the standard algorithm used for the ATLAS experiment. The default - the anti-k_T jet algorithm - cannot be used for this analysis: due to the reversed reconstruction scheme, soft particles are assigned to the jet in the last merging steps, resulting in unusable flip-values. The k_T algorithm in the exclusive mode, on the contrary, merges highly energetic objects at the last reconstruction steps. Therefore, the flip-values really describe the transition from 3 to 2 hard jets. In order to have access to jets reconstructed with this jet algorithm, the program ATHENA 15.6.10.6 has been used to re-run this algorithm on real data. After applying a good run list and a jet trigger, the data from run periods A to E have been cleaned from bad jets. To separate 3-jet-events from 4-jet-events, the impurity due to 4-jet-events has been calculated.

The real data has been compared to fully simulated dijet samples, generated by the Monte Carlo generator PYTHIA and also to calculations from the numerical integration program NLOJET++ (version 4.1.3). Data and simulations have shown a good agreement at the mass of the Z boson.

To cope with the influence of the jet-energy-scale (JES), a bin-by-bin correction has been applied. The correction factors have been obtained using PYTHIA, comparing jets from simulations with and without detector effects. As this method is not as precise as the measurement done for the anti-k_T jet algorithm with an already quite small JES uncertainty, a conservative estimate of 10% uncertainty has been considered for systematic studies of the JES.

α_s has then been determined by fitting the d_{23} distributions from real data to next-to-leading order (NLO) perturbative QCD predictions from calculations using the program NLOJET++. As NLOJET++ only calculates cross sections for parton productions, the influence of the hadronization has been adjusted bin-by-bin. The correction factors have been obtained via simulations using PYTHIA (version 6.4.24), comparing simulations on hadron and parton level. In order to measure the uncertainty of the hadronization, HERWIG (version 6.510) has been used as a comparison, as it has a different hadronization model implemented.

To take the influence of the Underlying Event (UE) into account, correction factors have been determined using PYTHIA, running hard scattering processes with and without the UE, using the UE tune AMBT1, which already includes ATLAS data. The calculations from NLOJET++ have then been adjusted bin-by-bin for the impact of the UE. For systematic studies, also the UE tunes ATLAS MC09c and PERUGIA10 have been studied.

Additionally, a method has been presented for correcting the influence of the UE directly on data. In this method, the transverse momenta of jets and their constituents are adjusted for the contribution of the UE using jets with a very small transverse momentum. It has been shown that this method can also be used to correct the d_{23} flip-values for the influence of the UE.

The fit from data to the calculations from NLOJET++ yielded a value of $\alpha_s(M_Z) = 0.120 \pm 0.001(stat.) \pm 0.005(syst.)$ at NLO, being in good agreement with the current world average. Systematic uncertainties due to the jet-energy-scale (± 0.004), the statistical uncertainty of the JES correction (± 0.001), the impurity due to 4-jet-events (± 0.002), the uncertainty of the renormalization scale (± 0.001), the PDF (± 0.001), the hadronization (± 0.001), the statistical uncertainty of the hadronization correction (± 0.001) and the uncertainty of the Underlying Event (± 0.001) have been studied.

As this method reproduces the results from former collider experements, this method can also be used to determine α_s in energy regimes not yet investigated.

Bibliography

[1] NAKAMURA, K. et al., *(Particle Data Group) Particle Physics Booklet.* (2010), J. Phys., G37:075021

[2] ERLEBACH, M., *Studien zu Multijet Ereignissen am Tevatron.* (2004), Diploma thesis, Ludwig-Maximilians-Universität München

[3] POVH, B., RITH, K., SCHOLZ, C. and ZETSCHE, F., *Teilchen und Kerne.* (2004), Springer-Verlag

[4] LICHTNECKER, M., *Studien zum Underlying Event und zu Pile-up beim ATLAS Experiment am LHC.* (2008), Diploma thesis, Ludwig-Maximilians-Universität München

[5] MAMEGHANI, R., *Semi- and Dileptonic Top Pair Decays at the ATLAS Experiment.* (2008), PhD thesis, Ludwig-Maximilians-Universität München

[6] LAMBACHER, M., *Study of full hadronic $t\bar{t}$ decays and their separation from QCD multijet background events in the first year of the ATLAS experiment.* (2007), PhD thesis, Ludwig-Maximilians-Universität München

[7] ENCYCLOPEDIA ENCARTA. (2003), Microsoft

[8] EGEDE, U., *The search for a standard model Higgs at the LHC and electron identification using transition radiation in the ATLAS tracker.* (1997), LUNFD6/(NFFL-7150)

[9] BARATE, R. et al., *Search for Standard Model Higgs Boson at LEP.* (2003), Phys. Lett. B, CERN-EP2003-011

[10] LEP, *http://public.web.cern.ch/public/en/research/lep-en.html.* (URL last visited: February 2011)

[11] TANO, V., *A study of QCD processes at low momentum transfer in hadron-hadron collisions.* (2001), PhD thesis, Rheinisch-Westfälische Technische Hochschule Aachen

[12] BIEBEL, O., *Experimental tests of the strong interaction and its energy dependence in electron-positron annihilation.* (2001), Physics Reports, Volume 340, Number 3, pp. 165-289(125)

[13] WEBBER, B. R., *Fragmentation and Hadronization.* (1999), arXiv:hep-ph/9912292v1

[14] DOKSHITZER, Y. L. and WEBBER, B. R., *Calculation of Power Corrections to Hadronic Event Shapes.* (1995), arXiv:hep-ph/9504219v1

[15] WEBBER, B. R., *Monte Carlo Methods in Particle Physics.* (2007), Talk at MPI Munich

[16] SJÖSTRAND, T. (1988), Int. J. Mod. Phys. A 3, 751

[17] NLOJET++-HOMEPAGE, *http://www.desy.de/~znagy/Site/NLOJet++.html.* (URL last visited: March 2011)

[18] DESY-HOMEPAGE, *http://www.desy.de/f/hera/engl/chap1.html*. (URL last visited: December 2010)

[19] HALZEN, F. and MARTIN, A., *Quarks and Leptons: An Introductory Course in Modern Particle Physics*. (1984), Wiley

[20] ELLIS, R. K., STIRLING, W. J. and WEBBER, B. R., *QCD and Collider Physics*. (2003), Cambridge University Press

[21] SJÖSTRAND, T., MRENNA, S. and SKANDS, P., *PYTHIA 6.4 Physics and Manual*. (2006), arXiv:hep-ph/0603175v2

[22] CTEQ META-PAGE, *http://www.phys.psu.edu/~cteq/*. (URL last visited: March 2011)

[23] BIEBEL, O., *QCD, Jets, Strukturfunktionen*. (WiSe 2003/2004), Vorlesungsreihe Angewandte Physik: Teilchenphysik mit höchstenergetischen Beschleunigern (TEVATRON und LHC), http://www.mppmu.mpg.de/

[24] DAWSON, I., BUTTAR, C. and MORAES, A., *Minimum Bias and the Underlying Event: towards the LHC*. (2003), Talk at Physics at LHC Conference in Prague

[25] REVOL, J.-P., *Diffractive Physics at the CERN Large Hadron Collider*. (2010), Talk at 6th International Conference on Physics and Astrophysics of Quark Gluon Plasma in Goa, India

[26] FIELD, R., *Min-Bias and the Underlying Event at Tevatron and the LHC, Lecture 1 and 2*. (2007), Talk at MCnet School Physics and Techniques of Event Generators at Durham University

[27] FERMILAB-HOMEPAGE, *http://www.fnal.gov/*. (URL last visited: January 2011)

[28] ATLAS-DATA-SUMMARY, *https://atlas.web.cern.ch/Atlas/GROUPS/DATAPREPARATION/ DataSummary/2010/records.html*. (URL last visited: March 2011)

[29] MORAES, A., *Minimum Bias interactions and the Underlying Event*. (2002), Talk at ATLAS UK Physics Meeting

[30] HIRSCHBÜHL, D. et al., *Measurement of the Charge Asymmetry in Top Pair Production*. (2007), http://www-cdf.fnal.gov/physics/new/top/2007/topProp/TopAsymmetry/

[31] ALBROW, M. et al., *Tevatron-for-LHC Report of the QCD Working Group*. (2006), arXiv:hep-ph/0610012v1

[32] SCHOUTEN, D., *Response Calibration and Pileup*. (2007), Talk at Trigger and Physics Week at CERN

[33] THE ATLAS COLLABORATION, *ATLAS Detector and Physics Performance – Technical Design Report, Volume I*. (1999), ATLAS TDR 14, CERN/LHCC/99-14

[34] THE ATLAS COLLABORATION, *The ATLAS Experiment at the CERN Large Hadron Collider*. (2008), JINST 3, S08003

[35] ATLAS-WIKI, *https://twiki.cern.ch/twiki/bin/view/Atlas/WorkBookAtlasExperiment*. (URL last visited: March 2011)

[36] CERN-HOMEPAGE, *www.cern.ch*. (URL last visited: December 2010)

[37] LICHTNECKER, F., *Atlas sculpture in front of the Rockefeller Center*. (Picture taken in New York on September 2007)

[38] ATLAS-WIKI, *https://twiki.cern.ch/twiki/bin/view/AtlasPublic/ AtlasTechnicalPaperListOfFigures.* (URL last visited: December 2010)

[39] LAMPRECHT, M., *Studien zu Effizienz und Akzeptanz des ATLAS-Myontriggers mit simulierten Messdaten.* (2007), Diploma thesis, Ludwig-Maximilians-Universität München

[40] LICHTNECKER, M., *End caps of the muon spectrometer.* (Picture taken at CERN on July 2006)

[41] ELSING, M. and SCHOERNER-SADENIUS, T., *Configuration of the ATLAS Trigger System.* (2003), physics/0306046

[42] GENEST, M.-H., *Jet triggers for early-data SUSY searches.* (2009), ATL-COM-PHYS-2009-437

[43] GUPTA, A., *Jet Algorithms in Athena.* (2003), ATLAS Physics Workshop in Athen

[44] MILLER, M. L., *Measurement of Jets and Jet Quenching at RHIC.* (2004), PHD thesis, Yale University

[45] BUTTERWORTH, J., COUCHMAN, J., COX, B. and WAUGH, B., *KtJet: A C++ Implementation of the k_T clustering algorithm.* (2002), arXiv:hep-ph/0210022v1

[46] CATANI, S., DOKSHITZER, Y. L., SEYMOUR, M. H. and WEBBER, B. R., *Longitudinally-invariant $k_\perp$ clustering algorithms for hadron-hadron collisions* (1993), Nucl. Phys. B406 187

[47] SCHMITT, C., *Study of the Underlying Event.* (2011), Talk at DPG Frühjahrstagung in Karlsruhe

[48] ATLAS-WIKI, *https://twiki.cern.ch/twiki/bin/view/AtlasProtected/Jetinputs?topic=JetInputs.* (URL last visited: January 2011)

[49] ATLAS-WIKI, *https://twiki.cern.ch/twiki/bin/view/AtlasProtected/CaloTowerJetInput.* (URL last visited: January 2011)

[50] TERASHI, K., *Jet and Missing ET Reconstruction and Signatures at ATLAS and CMS.* (2009), ATL-PHYS-PROC-2009-137

[51] ATLAS-WIKI, *https://twiki.cern.ch/twiki/bin/view/AtlasProtected/JetCalibration.* (URL last visited: January 2011)

[52] THE ATLAS COLLABORATION, *Update on the jet energy scale systematic uncertainty for jets produced in proton-proton collisions at $\sqrt{s}=7$ TeV measured with the ATLAS detector.* (2011), ATLAS-CONF-2011-007

[53] ATLAS-WIKI, *https://twiki.cern.ch/twiki/bin/view/AtlasProtected/HowToCleanJets.* (URL last visited: January 2011)

[54] THE ATLAS COLLABORATION, *Data-Quality Requirements and Event Cleaning for Jets and Missing Transverse Energy Reconstruction with the ATLAS Detector in Proton-Proton Collisions at a Center-of-Mass Energy of $\sqrt{s}=7$ TeV.* (2010), ATLAS-CONF-2010-038

[55] CORCELLA, G., KNOWLES, I. G. et al., *HERWIG 6.5: An event generator for Hadron Emission Reactions With Interfering Gluons (including supersymmetric processes).* (2001), JHEP, 0101:010

[56] ATLAS-WIKI, *https://twiki.cern.ch/twiki/bin/viewauth/Atlas/AthenaFramework.* (URL last visited: January 2011)

[57] CATANI, S. and SEYMOUR, M. H., *A Gerneral Algorithm for Calculating Jet Cross Sections in NLO QCD*. (1996), arXiv:hep-ph/9605323v3

[58] TRÓCSÁNYI, Z., *Progress in QCD next-to-leading order calculations*. (2002), arXiv:hep-ph/0201035v1

[59] ATLAS-WIKI, *https://twiki.cern.ch/twiki/bin/view/AtlasProtected/NLOJetForAtlas*. (URL last visited: January 2011)

[60] CACCIARI, M., SALAM, G. P. and SOYEZ, G., *FastJet 2.4.2 user manual*. (2010), http://www.lpthe.jussieu.fr/~salam/fastjet/

[61] LHAPDF, *http://projects.hepforge.org/lhapdf/*. (URL last visited: January 2011)

[62] THE ATLAS COLLABORATION, *ATLAS Monte Carlo tunes for MC09*. (2010), Tech. Rep. ATL-PHYS-PUB-2010-002

[63] SAMSET, B., *The latest results from ATLAS*. (2007), Talk at Pascos 2010 in Valencia

[64] SKANDS, P., *The Perugia Tunes*. (2009), arXiv:0905.3418v1 [hep-ph]

[65] FIELD, R., *Minimum Bias at CDF and CMS*. (2006), Talk at CMS Min-Bias Meeting

[66] BUTTAR, C. M., BUTTERWORTH, J. M. et al., *The Underlying Event*. (2006), www.desy.de/~heralhc/proceedings/WG2-Buttar.pdf

[67] SKANDS, P., *http://skands.web.cern.ch/skands/*. (URL last visited: March 2011)

[68] BUTTERWORTH, J., FORSHAW, J. et al., *JIMMY Generator Multiparton Interactions in HERWIG*. (URL last visited: April 2011), http://projects.hepforge.org/jimmy/

[69] PYTHON SOFTWARE FOUNDATION, *Python Programming Language*. (URL last visited: January 2011), http://www.python.org/

[70] THE GAUDI COLLABORATION, *http://proj-gaudi.web.cern.ch*. (URL last visited: January 2011)

[71] GEANT 4, *http://geant4.web.cern.ch/geant4/*. (URL last visited: January 2011)

[72] MANGANO, M. L., PICCININI, F. and POLOSA, A. D., *ALPGEN, a generator for hard multi-parton processes in hadronic collisions*. (2003), arXiv:hep-ph/0206293v2

[73] FRIXIONE, S. and WEBBER, B. R., *The MC@NLO 3.3 Event Generator*. (2006), arXiv:hep-ph/0612272v1

[74] ATLFAST, *http://www.hep.ucl.ac.uk/atlas/atlfast/*. (URL last visited: January 2011)

[75] ROOT-HOMEPAGE, *http://root.cern.ch/*. (URL last visited: January 2011)

[76] ATLAS-WIKI, *https://twiki.cern.ch/twiki/bin/view/AtlasProtected/JetConfigurableJobO*. (URL last visited: January 2011)

[77] ATLAS OFFLINE SOFTWARE RELEASE STATUS, *http://atlas-computing.web.cern.ch/atlas-computing/projects/releases/status/*. (URL last visited: March 2011)

[78] FREDERIX, R., *NLO QCD corrections to five-jet production at LEP and the extractions of $\alpha_s(M_Z)$*. (2010), arXiv:1008.5313v2 [hep-ph]

[79] PICH, A., *QCD Description of Hadronic Tau Decays*. (2011), arXiv:1101.2107v1 [hep-ph]

[80] DOKSHITZER, Y., *Contribution cited in Report of the Hard QCD Working Group, Proc. Workshop on Jet Studies at LEP and HERA*. (1990)

[81] BETHKE, S., *The 2009 World Average of α_s*. (2009), arXiv:0908.1135v2 [hep-ph]

[82] AMI, *https://atlastagcollector.in2p3.fr:8443/AMI/servlet/net.hep.atlas.Database.Bookkeeping.AMI.Servlet.Command?linkId=454*. (URL last visited: March 2011)

[83] ATLAS-WIKI, *https://twiki.cern.ch/twiki/bin/view/AtlasPublic/RunStatsPublicResults2010*. (URL last visited: March 2011)

[84] ATLAS-WIKI, *https://twiki.cern.ch/twiki/bin/view/AtlasProtected/SMJetAnalysis2010*. (URL last visited: January 2011)

[85] ANGERAMI, A. et al., *Methods for Background Subtraction for the kt and anti-kt Algorithms in Heavy Ion Events*. (2010)

[86] ALTARELLI, G. and PARISI, G., *Asymptotic Freedom in Parton Language*. (1977), Nucl. Phys., B126, 298

[87] GOTSMAN, E. et al., *Towards a new global QCD analysis: low x DIS data from non-linear evolution*. (2002), arXiv:hep-ph/0209074v1

[88] ANDERSEN, J. et al., *Small x Collaboration*. (2004), Eur. Phys. J. C 35, 67

[89] MÜLLER, T., *Investigation of the High Mass Drell Yan Spectrum with ATLAS*. (2010), PhD thesis, Ludwig-Maximilians-Universität München

[90] PESKIN, M. and SCHROEDER, D., *An Introduction to Quantum Field Theory*. (1995), Westview-Press

Die VDM Verlagsservicegesellschaft sucht für wissenschaftliche Verlage abgeschlossene und herausragende

Dissertationen, Habilitationen, Diplomarbeiten, Master Theses, Magisterarbeiten usw.

für die kostenlose Publikation als Fachbuch.

Sie verfügen über eine Arbeit, die hohen inhaltlichen und formalen Ansprüchen genügt, und haben Interesse an einer honorarvergüteten Publikation?

Dann senden Sie bitte erste Informationen über sich und Ihre Arbeit per Email an *info@vdm-vsg.de*.

Sie erhalten kurzfristig unser Feedback!

VDM Verlagsservicegesellschaft mbH
Dudweiler Landstr. 99
D - 66123 Saarbrücken
Telefon +49 681 3720 174
Fax +49 681 3720 1749
www.vdm-vsg.de

Die VDM Verlagsservicegesellschaft mbH vertritt

Printed by Books on Demand GmbH, Norderstedt / Germany